Hayk Sedrakyan
Nairi Sedrakyan

Math Kangaroo preparation book:
Level 5-6

2023

About the authors

Hayk Sedrakyan is an IMO medal winner, professional mathematical Olympiad coach in greater Boston area, USA. He has been the dean (and one of the main developers) of one of the biggest math competition preparation programs in the USA. He has been a Professor of mathematics in Paris and has a PhD in mathematics from the UPMC - Sorbonne University, Paris, France. Hayk is a Doctor of mathematical sciences in USA, France, Armenia and holds three master's degrees in mathematics from institutions in Germany, Austria, Armenia and has spent a small part of his PhD studies in Italy. Hayk Sedrakyan has worked as a scientific researcher for the European Commission (sadco project), has been one of the Team Leaders at Harvard-MIT Mathematics Tournament (HMMT) and was an invited speaker at Imperial College London. He took part in the International Mathematical Olympiads (IMO) in United Kingdom, Japan and Greece. Hayk has been elected as the President of the general assembly and a member of the management board of the *Cite Internationale Universitaire de Paris* (10,000 students, 162 different nationalities) and the same year they were nominated for the Nobel Prize. Hayk Sedrakyan is the son and was one of the students of Nairi Sedrakyan.

Nairi Sedrakyan is involved in national and international Mathematics Olympiads having been the President of Armenian Mathematics Olympiads and a member of the IMO problem selection committee. Nairi Sedrakyan was the winner of **Erdös Award** 2022 (one of the highest international awards in mathematics) for contributing to the development of mathematics worldwide, in average 1 person per year (globally) wins this award. He is one of 3 people from entire ex-USSR territory ever to win Erdös Award, the other 2 are: Grigori Perelman's professor (the only person in the world who coached 2 Fields Medal winners) and Nikolay Konstantinov (founder of Tournament of the Towns). Nairi Sedrakyan is the author of **the most difficult problem ever proposed in the history of the International Mathematical Olympiad (IMO)**, 5th problem of 37th IMO. This problem is considered to be the hardest problems ever in IMO because none of the members of the strongest teams (national Olympic teams of China, USA, Russia) succeeded to solve it correctly and because national Olympic team of China (the strongest team in IMO) obtained a cumulative result equal to 0 points and was ranked 6th in the final ranking of the countries instead of the usual 1st or 2nd place. The British 2014 film X+Y, released in the USA as *A Brilliant Young Mind*, inspired by the film *Beautiful Young Minds* (focuses on an English mathematical genius chosen to represent the United Kingdom at IMO) also states that this problem is the hardest problem ever proposed in the history of IMO (minutes 9:40-10:30). His students have received 20 medals in International Mathematical Olympiad, including Gold and Silver medals. Nairi Sedrakyan received a Gold Medal for contributions to World's Mathematical Olympiads and Scientific Activities.

Any comments or suggestions?
Then, please contact **sedrakyan.hayk@gmail.com**

Overview

This book consists of the **author-created new practice tests with author-prepared solutions (never published before)**. The authors provide detailed solutions to each problem and a list of answers. The main goal of the book is to teach problem solving strategies, how to solve non-standard problems and how to score better on **Math Kangaroo** (level 5-6) and other middle school math competitions. The book includes 210 new problems (7 practice tests and each practice test includes 30 problems). It is intended as a teacher's manual of mathematics, as a manual for math competition coaches, a self-study handbook for middle-school students and mathematical competitors.

Keywords: Math Kangaroo preparation, Competition math for middle school, American Mathematics Competitions preparation.

Any comments or suggestions?
Then, please contact **sedrakyan.hayk@gmail.com**

Mathematical competition is not about winning or losing, it is about mastering the art of thinking creatively and smart.

Hayk Sedrakyan.

Contents

Acknowledgment

The authors would like to thank their family for the support.

To Margarita, Ani, Jane and Luna.

The authors would like to thank *Jane Sukiasyan* for the cover and for the illustrations.

Test 1

Part A: Each correct answer is worth 3 points

1. Each row and each column of a 4×4 square consists of these four shapes ●, △, □, ○. In which of the squares a, b, c, d, e is the shape ○?

(A) a (B) b (C) c (D) d (E) e

2. There are four pieces of paper on the table. What is their correct order (from bottom to top)?

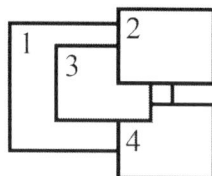

(A) 1, 2, 3, 4   (B) 2, 1, 3, 4   (C) 3, 2, 4, 1   (D) 4, 3, 2, 1   (E) 1, 4, 3, 2

3. The sum of the number of floors that are below and above Luna's apartment is 13. How many floors are there in Luna's building?

(A) 13   (B) 14   (C) 15   (D) 16   (E) 20

4. Two transparent paper squares are colored like this.

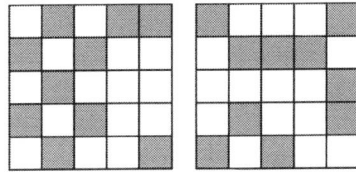

Placing these squares on each other, which of the following squares is not possible to get?

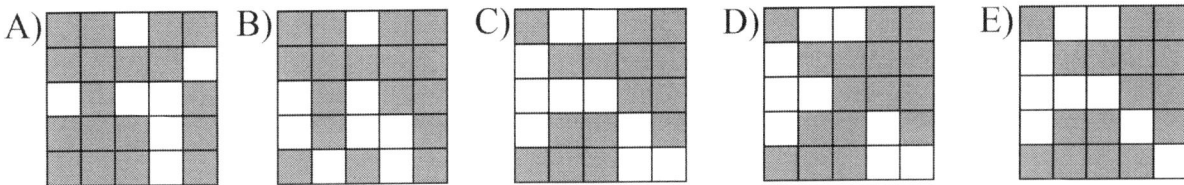

A)   B)   C)   D)   E)

5. In two places in between of the digits of the number 20245 plus (+) signs were placed, so that the sum is the smallest possible (out of all possible such sums). What is that sum?

(A) 43   (B) 49   (C) 67   (D) 211   (E) 247

6. How many digits are there, so that you can move one stick to get another digit?

$$0123456789$$

(A) 2   (B) 3   (C) 4   (D) 5   (E) 6

7. There are dozens of pens in the box. You are allowed to choose some of these pens and divide the chosen pens into two groups, so that the number of pens in each group is a prime number. Then, you need to move one or more pens from the first group to the second group, so that after this move the number of pens in each group is a prime number. What is the smallest number of pens that you can move so that all these conditions are true?

(A) 1   (B) 2   (C) 3   (D) 4   (E) 10

8. Which path from A to B is the shortest?

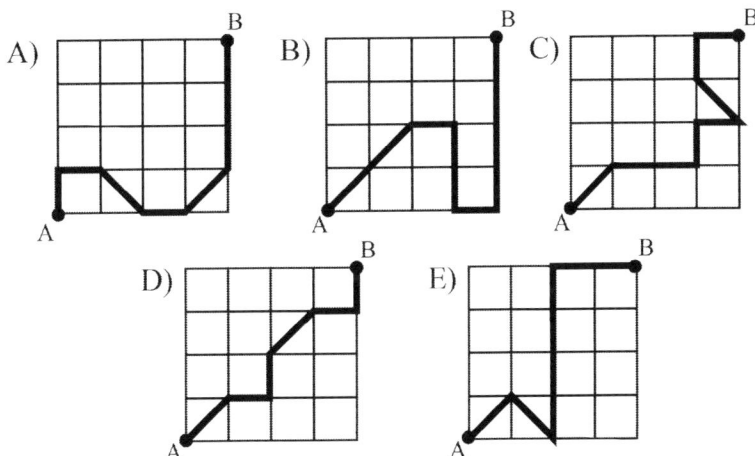

9. Which of these shapes is not possible to divide into two such shapes?

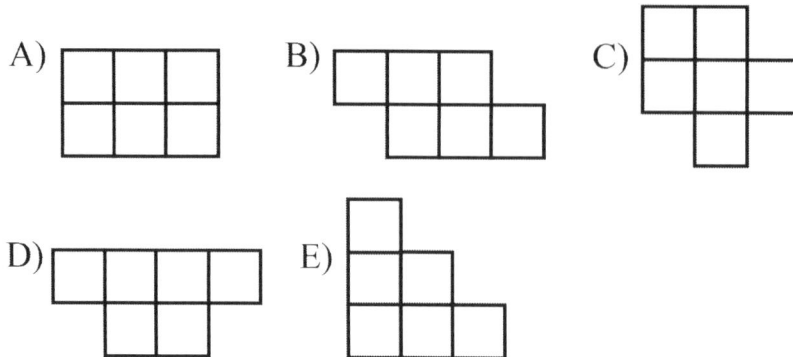

A)   B)   C)

D)   E)

10. What is the sum of the digits of the smallest natural number that is not divisible by any of its digits?

(A) 9 (B) 7 (C) 6 (D) 5 (E) 4

Part B: Each correct answer is worth 4 points

11. Phil is a darts player, at each turn he gets 10 out of 10, or 9 out of 10. Phil played 10 turns and he did not get the same score at any three turns in a row. What is Phil's smallest possible score after these 10 turns?

(A) 90 (B) 91 (C) 92 (D) 93 (E) 94

12. Luna's apartment is in one of the floors of a multi-storey building. The sum of the numbers of all other floors (except her floor) is 17. How many floors are there in the building?

(A) 4 (B) 5 (C) 6 (D) 7 (E) 8

13. A whole number n is equal to the sum of two two-digit multiples of 5. What is the greatest possible value of the sum of the digits of n?

(A) 10 (B) 12 (C) 13 (D) (E)

14. What is the area of the shaded shape? The area of each square is 1.

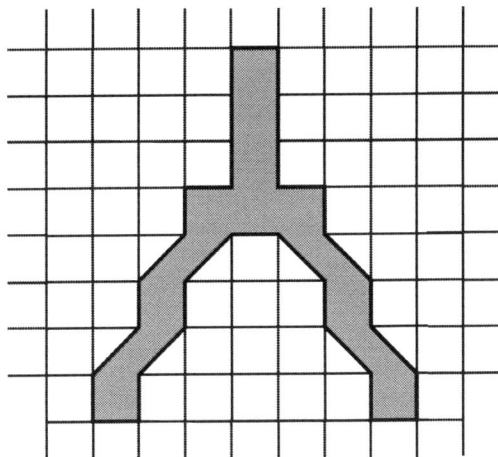

(A) 11.5   (B) 12   (C) 12.5   (D) 13   (E) 14

15. Three-sided football is a variation of football played with three teams instead of the usual two (see the figure). After each game only one team wins and the other two lose. 101 teams took part in a three-sided football tournament, where the losing team leaves. One of the teams won the tournament, how many games were played?

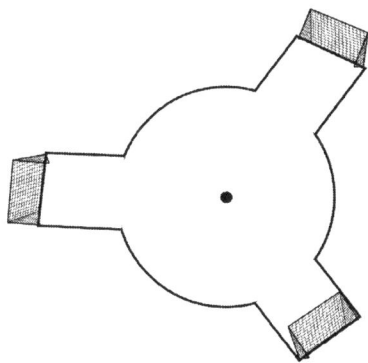

(A) 33   (B) 34   (C) 35   (D) 50   (E) 51

16. A three-digit number is 11 times greater than the sum of its digits. What is the product of its digits?

(A) 72   (B) 60   (C) 45   (D) 30   (E) 18

17. Using only one pair of parentheses in the expression $4 \cdot 2 + 20 : 4$ we can get different expressions. What is the greatest possible value among these expressions?

(A) 7   (B) 22   (C) 28   (D) 30   (E) 32

18. A, B, C, D are points on a plane, so that $AB = 3$, $BC = 30$, $AC = 33$, $AD = 23$, $BD = 20$. What is the length CD?

(A) 3   (B) 6   (C) 8   (D) 9   (E) 10

19. All these shapes consist of 12 cubes. Which of these shapes is not possible to divide into such shapes?

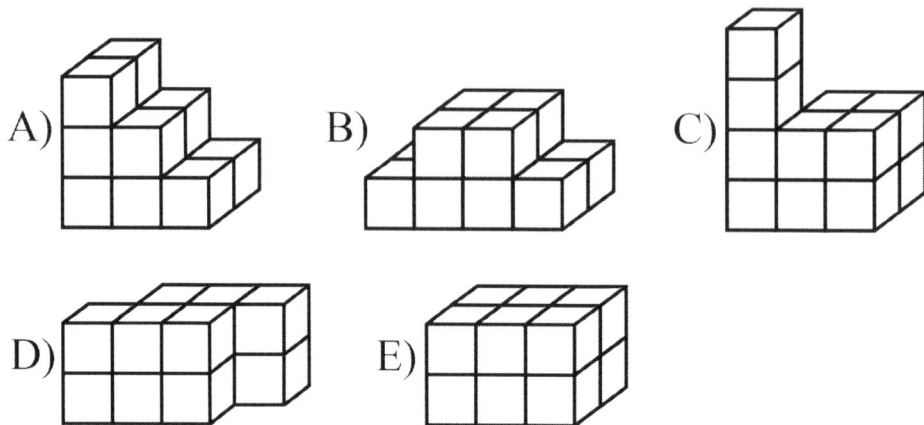

20. A little boy sitting next to the river noticed that two groups of ducks are swimming in the river. In the first group 13 ducks, in the second group 7 ducks. From time to time some ducks were moving from one group to the other one. After some time the number of ducks in both groups was the same. What is the greatest possible number of ducks, so that each of them made an odd number of moves from one group to another?

(A) 3   (B) 10   (C) 17   (D) 18   (E) 20

Part C: Each correct answer is worth 5 points

21. A scale is broken and gives a reading that is 10 percent less than the actual weight of the item. It shows that a bag of pears weights 5.4 pounds. What is the actual weight (in pounds) of this bag of pears?

(A) 4.86   (B) 5   (C) 5.1   (D) 5.15   (E) 6

22. A girl standing in a garden is 5 meters apart from a bird and she is 4 meters apart from a rabbit. At least how far apart can the bird and the rabbit be?

(A) 0.5 meters   (B) 0.8 meters   (C) 1 meter   (D) 2 meters   (E) 9 meters

23. The sum of two different positive whole numbers is six times their greatest common factor. What is the largest possible quotient when the largest of these numbers is divided by the smallest of these numbers?

(A) 2   (B) 3   (C) 4   (D) 5   (E) 10

24. A river fence consists of six vertical columns and some steel pickets. There are 9 steel pickets in part A, 12 steel pickets in part E. The difference of the number of pickets in any two neighboring parts is not more than 2. What is the greatest possible number of all pickets in parts B, C, D altogether?

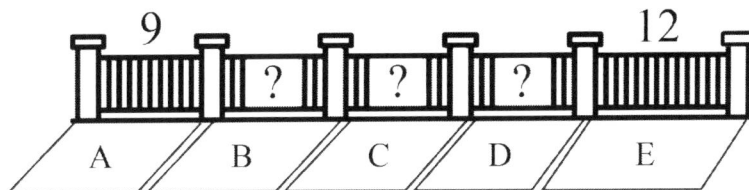

(A) 33   (B) 37   (C) 38   (D) 39   (E) 40

25. Ann has two dimes and three quarters. Bob has two dimes. What is the sum of all possible amounts (in cents) that Ann can pay to Bob?

(A) 950   (B) 860   (C) 855   (D) 780   (E) 740

26. Given $\angle ABC = 120°$. Each of rays BD and BE divide angle ABC into two angles, so that the ratio of their measures is $1 : 5$ (see the figure). What is the measure of $\angle DBE$?

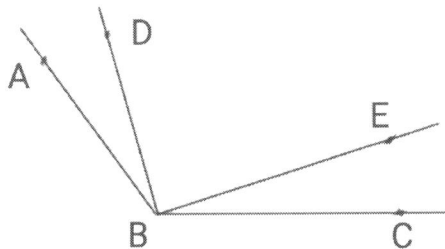

(A) 90°   (B) 80°   (C) 70°   (D) 65°   (E) 60°

27. Annie has several suitcases. There are two suitcases in one of them, there is one suitcase in two of them, and there is no suitcase in the remaining two suitcases. How many suitcases does Annie have?

(A) 3   (B) 4   (C) 5   (D) 6   (E) 7

13

28. River current speed is 2.4 meters per second. Moving with the river current ducks cross the bridge in 2 seconds, and moving against the river current ducks cross the bridge in 5 seconds. How wide is the bridge?

(A) 8 meters   (B) 16 meters   (C) 18 meters   (D) 14 meters   (E) 12 meters

29. In a shooting tournament Mia shot seven times and each time she got one of the following scores 0, 1, 2, ..., 10. Given that the sum of the scores of any three consecutive shots are different from each other. What is the greatest possible overall score that Mia could get after her seven shots?

(A) 64   (B) 65   (C) 66   (D) 67   (E) 68

30. At most how many squares of a 4×4 square can be painted, so that the painted squares do not form any such shape ?

(A) 8   (B) 9   (C) 10   (D) 11   (E) 12

Test 2

Part A: Each correct answer is worth 3 points

1. The international Morse code encodes the 26 letters of the English alphabet and all nine digits from 0 to 9 (see the figure).

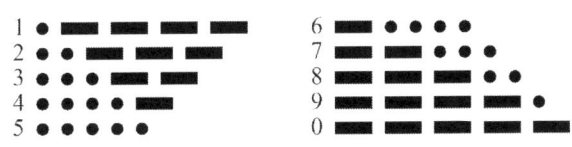

What is the value of this expression?

(A) 15   (B) 16   (C) 17   (D) 18   (E) 19

2. A garden is in the shape of a rectangle. 50 trees are planted 2 meters apart along its perimeter, so that there is one tree in every corner of the garden. What is its perimeter?

(A) 43 meters   (B) 50 meters   (C) 98 meters   (D) 100 meters   (E) 150 meters

3. Little Luna has built a tower using 7 green cubes and a few white cubes, so that the number of white cubes is less than 7. She did not put two cubes of the same color on each other. How many cubes are there?

(A) 8   (B) 10   (C) 13   (D) 14   (E) 15

4. Fill in each circle with an even digit, so that all inequalities are correct (see the figure).

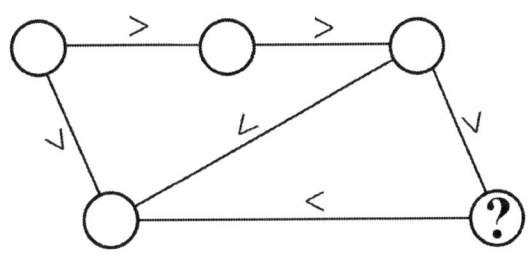

What digit must be in the rightmost circle?

(A) 0   (B) 2   (C) 4   (D) 6   (E) 8

5. For natural number n we know that

$$\frac{4}{5} + \frac{2}{7} = \frac{n}{70}.$$

What is n?

(A) 67   (B) 70   (C) 72   (D) 76   (E) 78

6. How many digits are there, so that you can move one stick to get another digit?

0123456789

(A) 0   (B) 1   (C) 2   (D) 3   (E) 4

7. How many of these shapes can be the common part of two triangles?

(A) 0   (B) 1   (C) 2   (D) 3   (E) 4

8. Annie has three nickels (5 cent coin), three dimes (10 cent coin), and three quarters (25 cent coin). At most, how many coins can she give to pay a bill of 90 cents?

(A) 5   (B) 6   (C) 7   (D) 8   (E) 9

9. A clown has green, red, and yellow balloons. Exactly six of them are not red and exactly four of them are not yellow. How many more yellow balloons are there than red balloons?

(A) 1   (B) 2   (C) 3   (D) 4   (E) 5

10. Given a square of side length 4. What is the area of the shaded part? (see the figure).

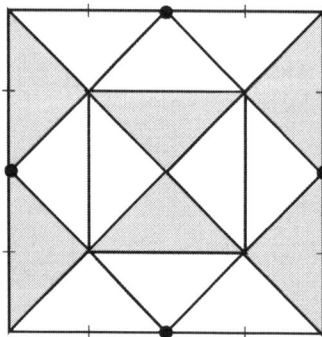

(A) 6   (B) 5   (C) 4   (D) 3   (E) 2.8

Part B: Each correct answer is worth 4 points

11. Mira drew three triangles (her triangles can have a common side). Which of the following numbers cannot be equal to the number of different sides of these triangles?

(A) 9   (B) 8   (C) 7   (D) 6   (E) 4

12. In how many different ways is it possible to create three pairs from the numbers 1, 2, 3, 4, 5, 6, so that the difference of the largest and the smallest numbers of each pair is a prime number?

(A) 0   (B) 1   (C) 2   (D) 3   (E) 4

13. Given two positive whole numbers, so that one of them is divisible by the other one. We know that the dividend is five times more than the divisor, and that the divisor is twice more than the quotient. What is the sum of the dividend, divisor, and quotient?

(A) 50   (B) 55   (C) 60   (D) 65   (E) 70

14. At most, how many common squares can a 4×8 rectangle and a 3×9 rectangle have?

(A) 12   (B) 16   (C) 24   (D) 26   (E) 27

15. Which of these shapes is not possible to form using eight cubes?

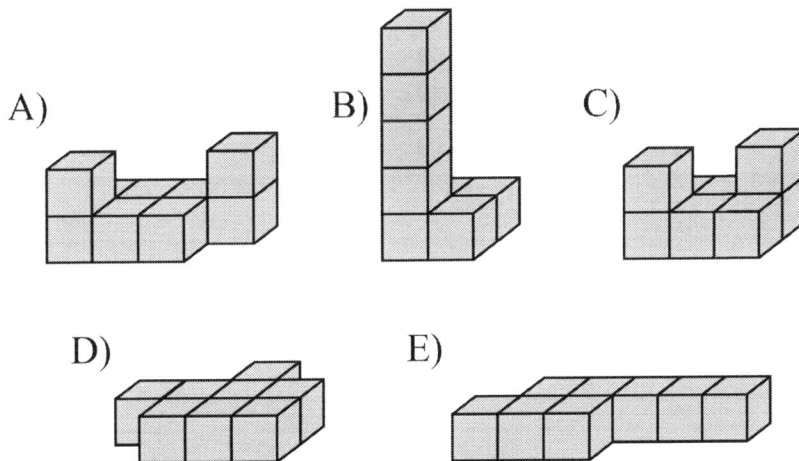

A)   B)   C)   D)   E)

16. Eight numbers are written in one line. The first number is 2, the second number is 3, then starting from the third number each number is equal to the greatest prime that is less than or equal to the sum of the previous two numbers. For example, the third number must be equal to 5, because the sum of the first two numbers is $2+3=5$ and the third number is equal to the greatest prime that is less than or equal to 5 (that prime is 5). What is the eighth number?

(A) 37   (B) 31   (C) 29   (D) 17   (E) 13

17. A cubical aquarium of edge length 6 centimeters is full of water. All the water is poured out into another cubical aquarium of edge length 8 centimeters. What is the height (in centimeters) of the water?

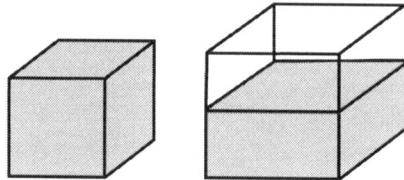

(A) 3   (B) 3.3   (C) 3.375   (D) 3.5   (E) 3.75

18. A pair (a, b) of natural numbers a and b is called "the simplest" if both $a - b$ and $a + b$ are prime numbers. If the pair (a, b) is "the simplest", then which of these numbers cannot be equal to $a - b$?

(A) 11   (B) 7   (C) 5   (D) 3   (E) 2

19. In every square of a 4×4 square is written one of the numbers 1, 2, 3, 4, so that in every row, in every column, and in every shaded part are written all four numbers 1, 2, 3, 4 (see the figure). What is $a + b$?

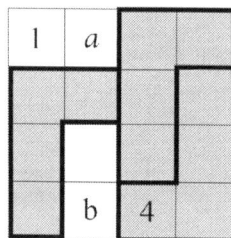

(A) 3   (B) 4   (C) 5   (D) 6   (E) 7

20. Four numbers are written on four pieces of paper (see the figure).

Putting together some of these paper pieces we can get new numbers. For example

How many numbers divisible by 15 can we get? You can use 1, 2, 3, or 4 pieces of paper.

(A) 4   (B) 6   (C) 8   (D) 10   (E) 11

Part C: Each correct answer is worth 5 points

21. A scale is broken and gives a reading that is 10 percent more than the actual weight of the item. It shows that a bag of pears weights 5.5 pounds. What is the actual weight (in pounds) of this bag of pears?

(A) 4.95 (B) 5 (C) 5.1 (D) 5.15 (E) 6

22. A 2×2 paper square needs to be cut into four shapes (see the figure), so that each of these four shapes includes the number 1 and the number 2. In how many different ways is it possible to do this?

		2	2
	1	1	
	1	1	
2	2		

(A) 0 (B) 1 (C) 2 (D) 3 (E) 4

23. Five-digit number $\overline{aabbb}$ is divisible by 32. What is the greatest possible value of $a + b$?

(A) 10 (B) 15 (C) 16 (D) 17 (E) 18

24. A 4×4 paper square and a 3×3 paper square each consist of 1×1 squares, the sides of which they can be cut along to form parts. What is the smallest possible number of parts needed to construct a 5×5 square?

(A) 3 (B) 4 (C) 5 (D) 6 (E) 7

25. A four-digit number is called "interesting" if exactly three of its digits are the same. How many four-digit "interesting" numbers are there so that for each of them the next number is also an "interesting" number?

(A) 8 (B) 17 (C) 72 (D) 73 (E) 81

26. A *cryptarithm* is a mathematical puzzle where the digits have been replaced by letters. How many solutions does the following cryptarithm have?

$$\overline{AR} \times \overline{TS} = \overline{AKH}.$$

(A) 0 (B) 1 (C) 2 (D) 3 (E) 4

27. Using exactly five of these six paper shapes a paper square was created (none of these shapes can have an overlapping region). Which shape was not used?

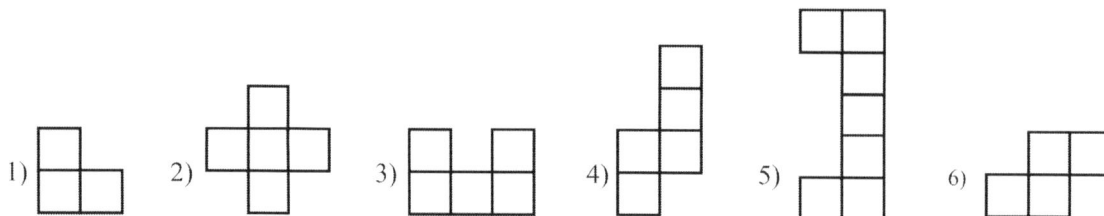

(A) 1　　　　(B) 2　　　　(C) 3　　　　(D) 5　　　　(E) 6

28. Two groups of ducks are swimming in the river. The number of ducks in the first group is a two-digit number $\overline{ab}$, the number of ducks in the second group is a two-digit number $\overline{ba}$. Some ducks moved from the first group to the second group, after this the number of ducks in the first group became a two-digit number $\overline{cd}$ and the number of ducks in the second group became a two-digit number $\overline{dc}$, where c and d may be the same, but $b \neq d$. At least how many ducks could move from the first group to the second group?

(A) 5　　　　(B) 6　　　　(C) 8　　　　(D) 9　　　　(E) 10

29. At least how many times a 6×6 square must be folded to get a 1×1 paper square?

(A) 5　　　　(B) 6　　　　(C) 7　　　　(D) 8　　　　(E) 9

30. The numbers 1, 2, 3, 4, 5, 6 are written on the faces of a cube (one number per face). For any two faces that share a common edge the positive difference of the numbers written on these two faces is written on that edge. What is the smallest possible sum of all numbers written on all 12 edges?

(A) 23　　　　(B) 24　　　　(C) 25　　　　(D) 26　　　　(E) 27

Test 3

1. Beach chairs and umbrellas are ordered in one line like this.

There are 50 umbrellas, at most how many chairs can there be?

(A) 50   (B) 52   (C) 100   (D) 101   (E) 102

2. Sam has two watches, one is 5 minutes behind and the other one is 10 minutes ahead. One of the watches shows that it is 10:27 am, the other one shows that it is 10:12 am. What time is it?

(A) 10:02 am   (B) 10:22 am   (C) 10:17 am   (D) 10:37 am   (E) cannot be determined

3. A paper rectangle was folded across a line parallel to one of its sides. Then, the folded paper was cut by a straight line. Which of these parts is not possible to get?

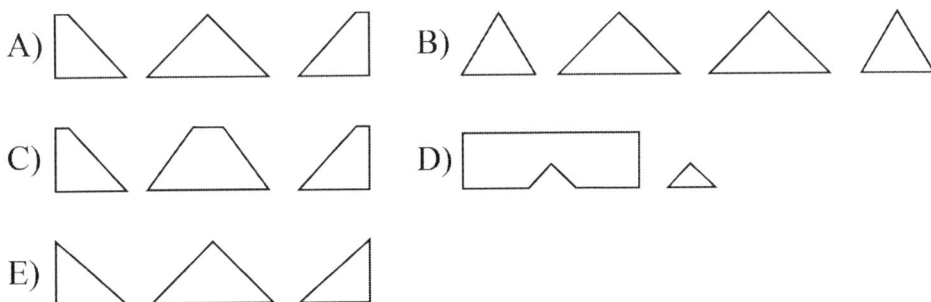

4. What is the sum of the digits of the smallest two-digit number $\overline{ab}$ that satisfies the following inequality $\overline{ab} < (a+1) \cdot (b+1)$?

(A) 7   (B) 8   (C) 9   (D) 10   (E) 18

5. At 11:30 am two groups of ducks were swimming in the river. Starting from noon till 12:17 pm, each 3 minutes 2 ducks moved from the first group to the second group, and every 5 minutes 3 ducks moved from the second group to the first group. Given that 12:16 pm there were 9 ducks in the first group. How many ducks were in the first group at 12:01 pm?

(A) 12   (B) 11   (C) 19   (D) 17   (E) 8

6. How many digits are there, so that there are at least two different ways to move one stick to get another digit?

$$0123456789$$

(A) 0   (B) 1   (C) 2   (D) 3   (E) 4

7. Each circle must be filled in by a different odd digit so that all inequalities are correct. What digit must be in the rightmost circle?

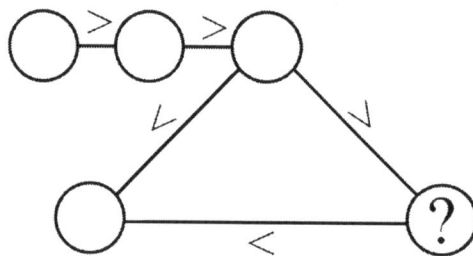

(A) 9   (B) 7   (C) 5   (D) 3   (E) 1

8. Five coins are placed in a row on a table. In how many different ways is possible to turn two coins? It is not allowed to turn any neighbor coins.

(A) 10   (B) 6   (C) 5   (D) 4   (E) 3

9. Six paper cards are placed in a row on a table. A number is written on each card, and three cards are face-up (as shown).

Given that the numbers on any three consecutive cards sum to the same number, what is the sum of the numbers on the three face-down cards?

(A) 21   (B) 17   (C) 16   (D) 12   (E) 10

10. How many different rectangular prisms is possible to construct using eight unit cubes?

(A) 1   (B) 2   (C) 3   (D) 4   (E) 6

Part B: Each correct answer is worth 4 points

11. The numbers 1, 1, 2, 2,..., 9, 9 are written on the blackboard. At most how many pairs of numbers is possible to form from these 18 numbers, so that the positive difference of the numbers in each pair is 1 or 2?

(A) 5   (B) 6   (C) 7   (D) 8   (E) 9

12. How many times 2 appears from 101 to 301?

(A) 20   (B) 40   (C) 100   (D) 120   (E) 140

13. Which of the following statements is correct?
(A) point M is outside the square and the triangle
(B) point M is outside the triangle and the circle
(C) point M is outside the circle or the square
(D) point M is inside the circle and the square
(E) point M is outside the circle and the square

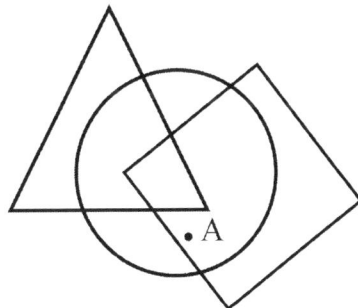

14. At least how many points need to be erased, so that no three leftover points are on the same line (see the figure)?

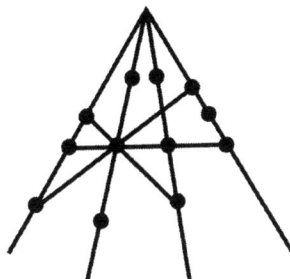

(A) 1   (B) 2   (C) 3   (D) 4   (E) 5

15. Mia needed to solve her math homework problems in six days. The first day she solved $\frac{1}{8}$ part of her homework, the next four days she solved three problems per day, the sixth day she solved the last two problems. How many problems did she solve the first day?

(A) 1   (B) 2   (C) 3   (D) 4   (E) 5

16. An elevator in a 9-story building has it's stops labeled on a digital screen as follows:

Mia was looking at the elevator's mirror and looked at the reflection of the screen and saw the number of the floor where she lives. When she left the elevator, she realized she had gotten off too early. On what floor does Mia live?

(A) 1   (B) 2   (C) 3   (D) 5   (E) 8

17. On the first day a car travelled $\frac{1}{3}$ part of the distance on a straight road, on the second day it travelled $\frac{1}{3}$ part of the remaining road, on the third day it travelled $\frac{1}{3}$ part of the remaining road, and on the fourth day it travelled all the remaining part of the road. What part of the entire road was travelled on the fourth day?

(A) $\frac{1}{3}$ (B) $\frac{2}{9}$ (C) $\frac{4}{27}$ (D) $\frac{8}{27}$ (E) $\frac{1}{4}$

18. A pentagon is divided by one of its diagonals into a triangle and a quadrilateral. The perimeters of a triangle, quadrilateral, and pentagon are 13, 14, 15, respectively. What is the length of that diagonal?

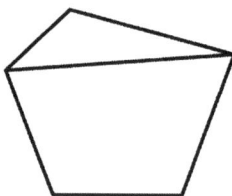

(A) 5   (B) 6   (C) 6.5   (D) 7   (E) 7.5

19. Sam and Luna had the same math homework problems to solve. Each day Sam solved three problems, and Luna solved four problems. When Luna was done with her entire math homework Sam needed to work four more days to be done. From how many problems did the math homework consist of?

(A) 20   (B) 24   (C) 36   (D) 48   (E) 68

20. Seven coins are placed in a row on a table. In how many different ways is possible to turn three coins? It is not allowed to turn any neighbor coins.

(A) 7   (B) 8   (C) 9   (D) 10   (E) 11

Part C: Each correct answer is worth 5 points

21. How many four-digit numbers are there that can be written using each of the following digits 1, 2, 3, 4, so that any two neighbor digits are of different parity?

(A) 16   (B) 20   (C) 24   (D) 32   (E) 64

22. There are apples and pears in a fruit bowl. The number of apples is 40 percent of all apples and pears in the bowl. Mia ate one of the apples and the number of apples became equal to 50 percent of all pears in the bowl. Then, Bob ate one of the pears. What percent of the pears are the apples?

(A) 60   (B) 55   (C) 50   (D) 40   (E) 33

23. There are 10 chairs next to a round table. At least how many people need to sit on these chairs (each chair can be used only by one person), so that for each chair at least one of its neighbor chairs is occupied by someone?

(A) 3   (B) 4   (C) 5   (D) 6   (E) 7

24. The common part of a rectangle of perimeter 6 and a rectangle of perimeter 8 is a rectangle of a perimeter 3 (see the figure). What is the perimeter of rectangle $PQRS$?

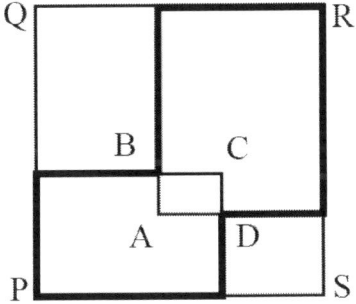

(A) 7   (B) 9   (C) 10   (D) 11   (E) 12

25. A floor is tiled with 20×40 tiles. An ant starts to move from point A to point B, it can move only along the sides of the tiles and it chooses the shortest route. In how many different ways can the ant go from A to B?

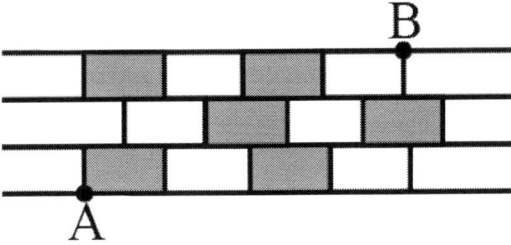

(A) 8   (B) 10   (C) 13   (D) 19   (E) 20

26. Luna needs to fry 5 hamburger patties. It takes 2 minutes to fry each side of each patty. At most 4 patties fit to her kitchen pan. At least in how many minutes can Luna fry all 5 patties?

(A) 4   (B) 5   (C) 6   (D) 7   (E) 8

27. How many five-digit numbers $\overline{abcde}$ are there, so that $b = a + 1$, $c = b + 1$, $e = d + 1$?

(A) 63   (B) 56   (C) 50   (D) 16   (E) 10

28. At most in how many squares of a 8×8 square is possible to put ★, so that there is at most one ★ in any such ⌐ shape (in any position: rotated or flipped)?

(A) 8   (B) 9   (C) 10   (D) 11   (E) 12

29. At least in how many colors can the squares of 8×8 square be painted, so that any shape (in any position: rotated or flipped) does not include two or more squares of the same color?

(A) 7   (B) 8   (C) 9   (D) 10   (E) 11

30. In how many different ways can we fill in the squares by the numbers 1, 2, 3, 4, 5, 6, 7, so that the sum of all four numbers of each circle is the same number? Only one number can be written in a square and all numbers must be used.

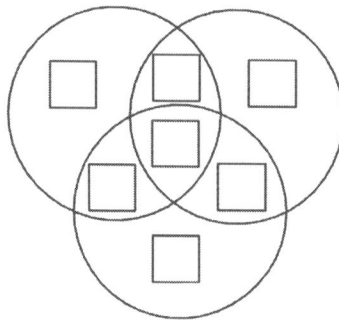

(A) 60   (B) 100   (C) 110   (D) 108   (E) 120

Test 4

1. A math teacher gave students five problems. Each student solved either three, four, or five problems. Two students solved five problems. The number of students who solved four problems is 3 times more than the number of students who solved five problems and is 2 times less than the number of students who solved three problems. How many students are there?

(A) 18 (B) 19 (C) 20 (D) 22 (E) 24

2. For example, a rectangle has two *lines of symmetry* (see the picture).

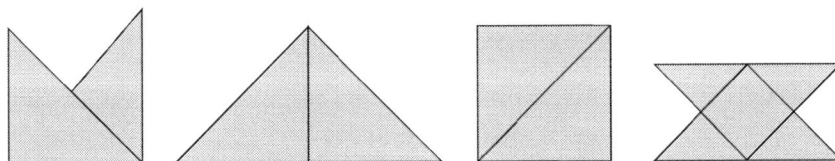

Match the number of lines of symmetry to each shape. Which answer is extra?

(A) 0 (B) 1 (C) 2 (D) 3 (E) 4

3. In a five-day school week a student needed to solve in average six problems per day. The student solved seven math problems in the first day, seven math problems in the second day, four math problems in the third day, and five math problems in the fourth day. How many problems are left to be solved in the fifth day?

(A) 4 (B) 5 (C) 6 (D) 7 (E) 8

4. The age of a girl in months is equal to the age of her grandmother in years. If the sum of their ages is 65 years, what is the (positive) age difference in years between them?

(A) 40 (B) 45 (C) 50 (D) 55 (E) 56

5. At most, how many of these five symbols $=, <, >, \leq, \geq$ can we use instead of ♣?

$$\frac{1}{2} + \frac{1}{3} + \frac{1}{5} + \frac{1}{6} \; \clubsuit \; 1 + \frac{1}{5}.$$

(A) 1 (B) 2 (C) 3 (D) 4 (E) 5

6. Which of these rectangular prisms can be constructed using at most 17 unit cubes?

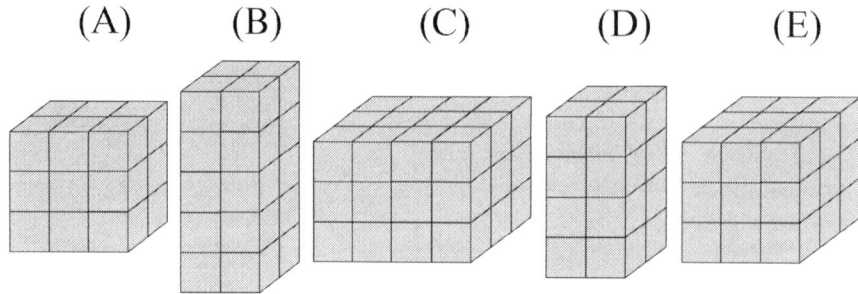

(A)   (B)   (C)   (D)   (E)

7. The product of three different natural numbers is equal to 30. Which of the following answers cannot be equal to the sum of these three numbers?

(A) 10   (B) 12   (C) 14   (D) 16   (E) 18

8. A natural number is called "successful" if its digits are consequent numbers in increasing order from left to right. For example, 45 and 567 are "successful" numbers. M is the smallest "successful" number greater than 100 and divisible by 67. What is the sum of the digits of M?

(A) 14   (B) 15   (C) 16   (D) 18   (E) 19

9. The following paper shape can be cut only into rectangles. At least, how many cut-outs (rectangle pieces) can be there?

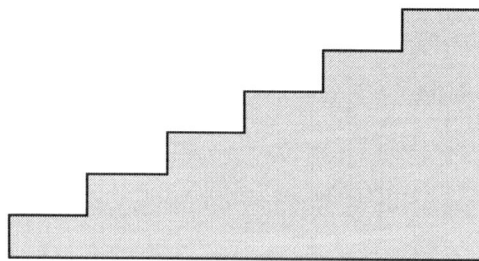

(A) 5   (B) 6   (C) 7   (D) 12   (E) 15

10. What is the number of all four-digit numbers that can be written using the digits 1, 2, 3, 4, so that each of these digits is used exactly once and the first digit is not 1, the second digit is not 2, the third digit is not 3, and the fourth digit is not 4?

(A) 6   (B) 9   (C) 10   (D) 12   (E) 20

Part B: Each correct answer is worth 4 points

11. Ann has some amount of money. If she decides to buy 11 copies of the same book, then she will have 50 cents left. If she decides to buy 15 such books, then she will need 70 cents more than she has. At least, how much more money (in cents) does Ann need to be able to buy 20 such books?

(A) 600   (B) 500   (C) 220   (D) 200   (E) 180

12. A square paper was folded into half twice, then a small part of it was cut out.

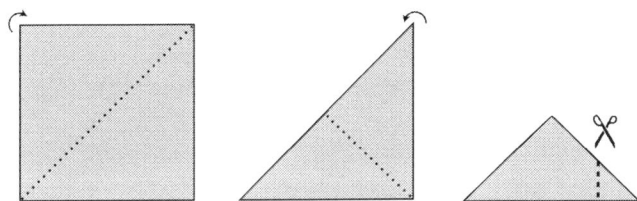

If we unfold it, which of the following shapes can we get?

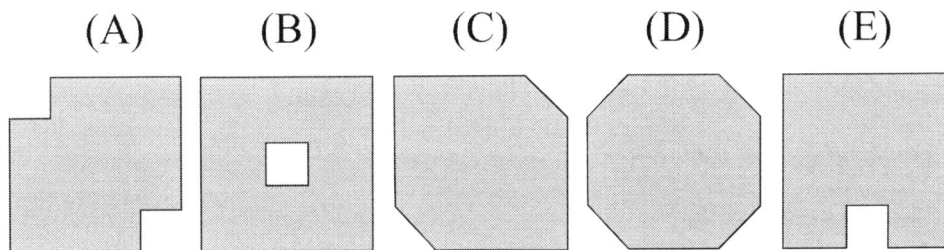

(A)   (B)   (C)   (D)   (E)

13. The height of each shelf of a kitchen rack is 25.5 cm. If we put four plates on top of each other their height is 6 cm. If we put seven such plates on top of each other their height is 9 cm.

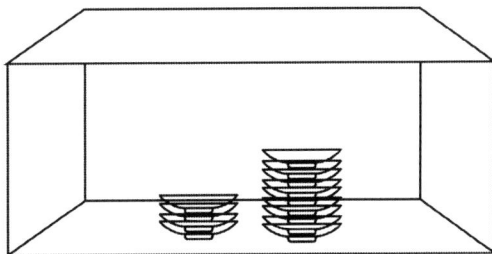

At most, how many plates can we put on top of each other in each shelf of this kitchen rack?

(A) 22   (B) 23   (C) 24   (D) 25   (E) 26

14. How many times does the number 5 appear from 1 to 200?

(A) 35   (B) 38   (C) 39   (D) 40   (E) 41

29

15. Given one large and four small rectangles (see the picture). Points A, B, C, D are the intersection points of the diagonals of each small rectangle. The area of the large rectangle is 20 and the area of each small rectangle is 2. What is the area of the shaded shape?

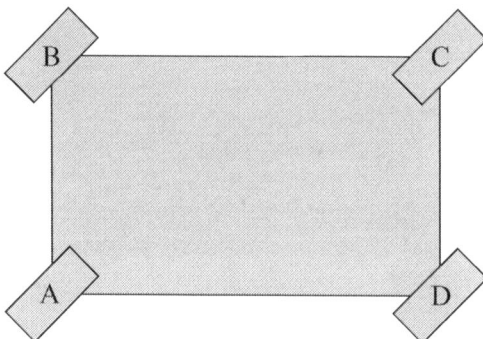

(A) 20   (B) 22   (C) 24   (D) 26   (E) 28

16. A picket fence consists of 60 wooden pickets (see the picture). Each wooden picket needs to be colored either in red, blue, or orange, so that any three consecutive pickets have different colors. In how many different ways is it possible to color this picket fence?

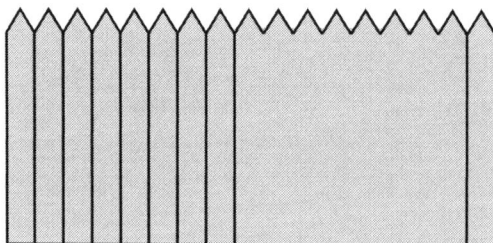

(A) 3^{60} (B) 81   (C) 27   (D) 18   (E) 6

17. In how many different ways can you draw a line from the top leftmost letter "k" to the bottom rightmost letter "k" to get the word "kayak"?

k	a	y
a	y	a
y	a	k

(A) 9   (B) 12   (C) 10   (D) 6   (E) 8

18. This shape consists of unit cubes (see the picture). How many unit squares make up the surface of this shape?

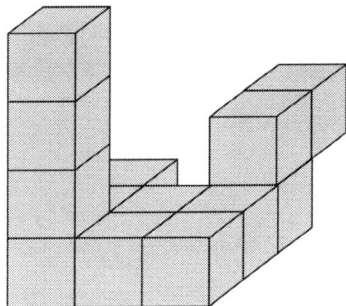

(A) 50   (B) 51   (C) 52   (D) 53   (E) 54

19. 1, 1, 1, 1, 2, 2, 2, 2, 3 are the side lengths of three triangles. What is the product of the perimeters of these three triangles?

(A) 105   (B) 120   (C) 140   (D) 150   (E) 180

20. A floor has a shape of a regular hexagon of side length 10 meters (see the picture). At least, how many equilateral triangle tiles of side length 1 decimeter are needed to tile this floor?

(A) 400   (B) 600   (C) 6000   (D) 10000   (E) 60000

Part C: Each correct answer is worth 5 points

21. A three-digit number is called an "interesting" number, if the sum of its digits is a prime number and the sum of any two digits is a prime number. How many three-digit "interesting" numbers are there?

(A) 8   (B) 9   (C) 13   (D) 17   (E) 18

22. The airplane travels the distance between two cities in 3 hours 20 minutes. If the airplane increases its speed by 200 kilometers per hour, then it travels the same distance in 2 hours 30 minutes. What is the distance (in kilometers) between these two cities?

(A) 500   (B) 600   (C) 1000   (D) 2000   (E) 2100

23. What is the sum of the digits of the smallest five-digit number so that none of its digits is divisible by each other?

(A) 30   (B) 31   (C) 32   (D) 33   (E) 34

24. How many numbers with different digits are there that are greater than 2023 and smaller than 2320?

(A) 110   (B) 119   (C) 200   (D) 204   (E) 225

25. What is the value of the following expression?

$$1011 - 1213 + 1415 - 1516 + 1617 - 1718 + 1819 - 2021 + \ldots + -9697 + 9899.$$

(A) 5555   (B) 5554   (C) 5553   (D) 5455   (E) 555

26. What is the number of all two-digit numbers $\overline{ab}$, so that $\overline{ba}$ is also a two-digit number and when $\overline{ab}$ is divided by $\overline{ba}$ the remainder is $a + b$?

(A) 1   (B) 2   (C) 3   (D) 5   (E) 7

27. There are 17 tables in a restaurant. The following placements are possible: tables that are placed separately, tables that are placed in pairs, tables that are placed in triplets. There are four chairs next to each table placed separately, there are six chairs next to each table placed in pairs, and there are eight chairs next to each table placed in triplets. Given that altogether there are 50 chairs. What is the difference of the number of tables placed in triplets and the number of tables placed separately?

(A) 1   (B) 2   (C) 3   (D) 4   (E) 5

28. At most, how many such shapes (in any position: rotated or flipped) can be cut out from a 7×9 paper rectangle? Each of these shapes must consist of four unit squares.

(A) 10   (B) 11   (C) 12   (D) 13   (E) 14

29. A 4×4 square consists of 16 unit squares. At least how many sides of these 16 squares must we erase, so that each square consisting of the sides of these unit squares includes at least one erased side?

(A) 6   (B) 9   (C) 10   (D) 5   (E) 8

30. The numbers 1, 2,..., 9 are written in the squares of a 3×3 square (one number per square). Consider all three row sums and all three column sums. At most, how many of these six numbers can be prime numbers?

(A) 2   (B) 3   (C) 4   (D) 5   (E) 6

Test 5

1. To make the equation $2050 + \bigstar - 1 = 2051$ correct, by which of the following must we replace $\bigstar$?

(A) 0   (B) 1   (C) 2   (D) 2050   (E) 2051

2. The sum of all six numbers written at the vertices of each hexagon is equal to 15 (see the picture).

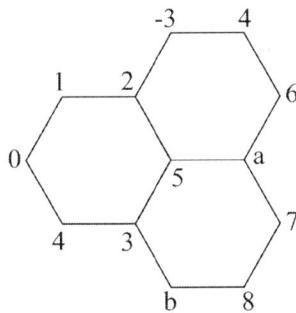

What is the value of $a - b$?

(A) 4   (B) 6   (C) 8   (D) 10   (E) 12

3. What is the sum of all whole numbers in between -12.01 and 14.03?

(A) 14   (B) 27   (C) 30   (D) 31   (E) 32

4. How many of the numbers 0, 1, 2, 3, 4, 5, 6, 7, 8, 9 have an *axis of symmetry*?

(A) 2   (B) 1   (C) 5   (D) 3   (E) 4

5. Given that

$$\frac{5}{6} + \frac{7}{50} = \frac{m}{300}.$$

What is the sum of the digits of m?

(A) 10   (B) 12   (C) 13   (D) 15   (E) 17

6. There are some red, blue, and black pens on a table. Exactly two of them are not black and exactly three of them are not red. How many black pens are there on the table?

(A) 1   (B) 2   (C) 3   (D) 4   (E) 5

7. A paper square was cut into one square of size 2×2 and n rectangles of size 1×3, where n is a natural number. What is the smallest possible value of n?

(A) 2   (B) 3   (C) 4   (D) 5   (E) 7

8. At most how many different digits must be used in order to write three consecutive three-digit numbers?

(A) 5   (B) 6   (C) 7   (D) 8   (E) 9

9. A digital clock is broken and does not show : symbol. So, instead of showing 17 : 30 it shows 1730. At most, how many different four-digit numbers can it show in one day?

(A) 840   (B) 841   (C) 960   (D) 1200   (E) 1440

10. Eight identical rectangles form a large rectangle (see the figure). What is the ratio of the length to the width of the large rectangle?

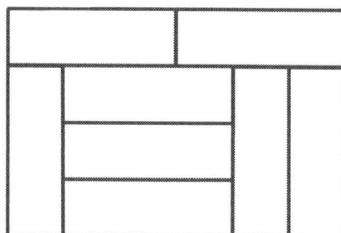

(A) 3:1   (B) 4:1   (C) 4:3   (D) 3:2   (E) 5:3

Part B: Each correct answer is worth 4 points

11. A chocolate bar consists of 32 squares. One almond is placed on any two squares that have a common side, so that half of the almond is in one square and the other half is in the other square (see the picture). How many almonds are used?

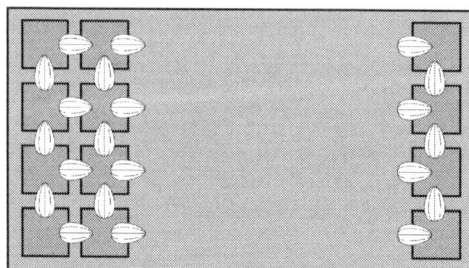

(A) 21   (B) 24   (C) 40   (D) 50   (E) 52

12. 12 chairs are placed next to a round table. At least two people sit on these chairs, each chair can be used only by one person. For any two people, at least one of them has two neighbors (is sitting in between of these two neighbors). How many people are there?

(A) 6   (B) 8   (C) 6 or 8   (D) cannot be determined   (E) other answer

13. At least, by how many straight lines must this paper shape be cut so that putting the pieces together we can make a hexagon?

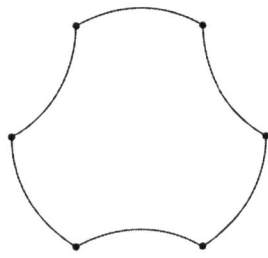

(A) 1   (B) 2   (C) 3   (D) 4   (E) 6

14. A big rectangle is divided into five rectangles (see the picture). The sum of the perimeters of four corner rectangles is 2024. What is the perimeter of the big rectangle?

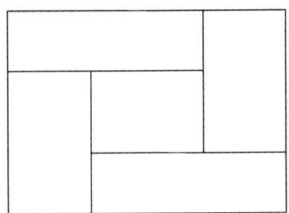

(A) 2024   (B) 1012   (C) 1000   (D) 4048   (E) 2000

15. What is the sum of the digits of the smallest natural number that the product of its digits is equal to 648?

(A) 12   (B) 15   (C) 16   (D) 25   (E) 26

16. The sum of three different natural numbers is 10. Which of the following numbers cannot be equal to their product?

(A) 14   (B) 16   (C) 18   (D) 20   (E) 30

17. In how many different ways is it possible to connect the letters to form the word AMERICA (see the picture)? Each letter can be connected only with its neighbor letters.

A	M	E	R
M	E	R	I
E	R	I	C
R	I	C	A

(A) 10   (B) 20   (C) 21   (D) 25   (E) 50

18. 30 students sit in 3 rows. Half of the students from the third row moved to the first two rows and the number of students in each of the first two rows doubled. How many students were in the third row?

(A) 10   (B) 14   (C) 18   (D) 20   (E) 24

19. At most, how many such shapes (in any position: rotated or flipped) is possible to cut out from a 5×5 paper square? Each shape must consist of three squares of size 1×1.

(A) 5   (B) 6   (C) 7   (D) 8   (E) 9

20. In each square of a 3×3 square is written a number. At least how many different numbers must be written in this 3×3 square, so that all six row and column sums are different from each other? (Row sum is the sum of three numbers in that row, column sum is the sum of three numbers in that column. There are three row sums and three column sums, so six sums).

(A) 1   (B) 2   (C) 3   (D) 4   (E) 5

Part C: Each correct answer is worth 5 points

21. An *Euler path* is a walk along the edges of this shape (of an edge length 1) which uses every edge exactly once. What is the length of the longest *Euler path* from A to B?

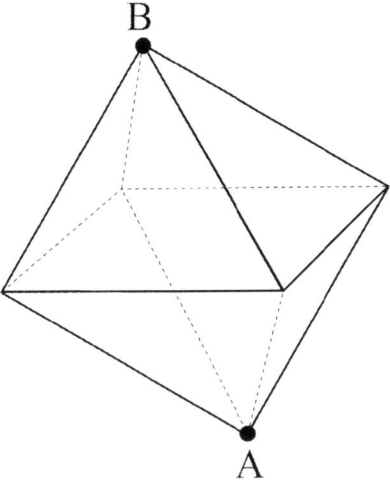

(A) 12   (B) 11   (C) 10   (D) 9   (E) 8

22. A natural number was written in each square of a 4×4 square. Some of these numbers were erased (see the picture). We know that the products of all four numbers of any row (column) are equal and can be written as a product of three prime numbers. What is a?

			a
2			
3	2		
5	3	2	

(A) 1   (B) 2   (C) 3   (D) 4   (E) 5

23. Winnie the pooh can eat one honey jar in 3 hours. Piglet can eat the same honey jar in 6 hours. In how many hours can they both together eat one honey jar?

(A) 2.5   (B) 2   (C) 1.5   (D) 1   (E) 0.5

24. How many nine-digit numbers are there with the digits in decreasing order? For example, 976543210 is such a nine-digit number.

(A) 3   (B) 5   (C) 8   (D) 9   (E) 10

25. Given a 3×3 square, where $a > 75 > b$. Consider all six row and column sums. Given that two of these sums are equal. What is $a + b$?

a	a	a
a	b	b
75	75	a

(A) 75   (B) 150   (C) 200   (D) 210   (E) 250

26. The sum of two-digit numbers $\overline{ab}$, $\overline{bc}$, $\overline{ca}$ is a perfect square. What is the value of $a + b + c$?

(A) 9   (B) 10   (C) 11   (D) 12   (E) 24

27. How many five-digit numbers are there with these two properties?
• Sum of its digits is 2. For example, 10001, as $1 + 0 + 0 + 0 + 1 = 2$.
• There is a natural number, so that when it is added to the five-digit number, then the sum of the digits of their sum is also 2. For example, for 10001 such natural number is 9, as $10001 + 9 = 10010$ and $1 + 0 + 0 + 1 + 0 = 2$.

(A) 0   (B) 1   (C) 2   (D) 3   (E) 4

28. In how many different ways is it possible to choose three numbers from 1, 2,..., 30, so that their product is 2024 and they are in increasing order? For example $8 \cdot 11 \cdot 23 = 2024$.

(A) 1   (B) 4   (C) 3   (D) 2   (E) 5

29. A 3×3 square consists of nine squares. At least how many sides of these nine squares must be removed, so that each square has at least one side that was removed?

(A) 3   (B) 4   (C) 5   (D) 6   (E) 7

30. Using three different digits it is possible to write six different three-digit numbers. For example, if we use the digits 1, 2, 3, we can write 123, 132, 213, 231, 312, 321. If we use the digits 2, 3, 4, we can write 234, 243, 324, 342, 423, 432. Some three different digits were used to write six different three-digit numbers, so that the sum of five of them is 2026. What is the product of the digits of the sixth three-digit number?

(A) 20   (B) 21   (C) 24   (D) 30   (E) 36

Test 6

1. Which of the following expressions is equal to 1?

(A) $1000 - \frac{997}{3}$ (B) $\frac{1000}{3} - 997$ (C) $\frac{1000}{3} - \frac{997}{3}$ (D) $997 - \frac{1000}{3}$ (E) $\frac{997}{3} - \frac{1000}{3}$

2. A wire was folded twice (see the picture). How long is this wire?

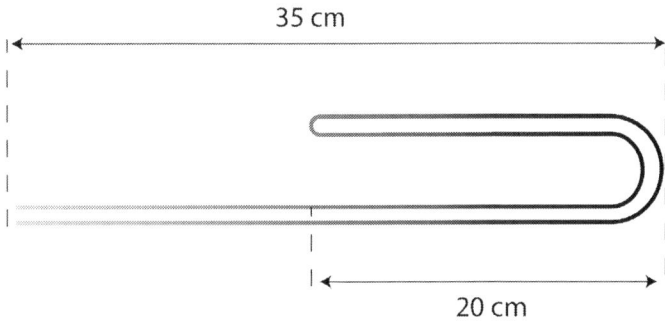

(A) 100 cm   (B) 110 cm   (C) 115 cm   (D) 120 cm   (E) 125 cm

3. The average of two numbers is 10. If one of the numbers is 12, what is the other number?

(A) 10   (B) 6   (C) 9   (D) 12   (E) 8

4. Given the points X(5), Y(4), Z(7) on the number line (see the picture). What is their correct order from left to right?

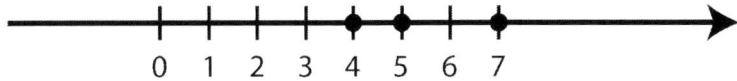

(A) X, Y, Z   (B) X, Z, Y   (C) Y, X, Z   (D) Y, Z, X   (E) Z, X, Y

5. Some Australian kangaroos can jump either 3 meters or 5 meters. At least, how many jumps are needed to travel a distance of exactly 48 meters?

(A) 16   (B) 12   (C) 11   (D) 9   (E) 10

6. Given two 1×3 paper rectangles, one black and one white. Which of the following shapes is not possible to get from these two paper rectangles? You cannot fold the rectangles.

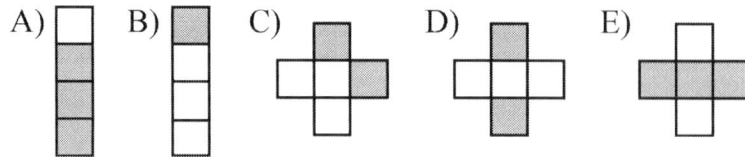

(A)   (B)   (C)   (D)   (E)

7. A digital clock shows the digits like this:

The clock is broken and does not show the middle horizontal line. Using the visible part of the digits we want to draw as many digits as possible. At most, how many digits can you definitely draw correctly?

(A) 10   (B) 9   (C) 8   (D) 7   (E) 6

8. Which of the following can be equal to the sum of five consecutive whole numbers?

(A) 101   (B) 102   (C) 103   (D) 104   (E) 105

9. Without calculating, which of the following products is the greatest?

$$2020 \cdot 2024, \quad 2021 \cdot 2023, \quad 2019 \cdot 2025, \quad 2018 \cdot 2026, \quad 2022 \cdot 2022.$$

(A) $2020 \cdot 2024$ (B) $2021 \cdot 2023$ (C) $2019 \cdot 2025$ (D) $2018 \cdot 2026$ (E) $2022 \cdot 2022$

10. Given $a - b = 2023$, where a and b are whole numbers. What is the smallest possible non-negative value of $a + b$?

(A) 0   (B) 1   (C) 20   (D) 100   (E) 2023

Part B: Each correct answer is worth 4 points

11. The screen of a mobile phone is a 6.5×11 rectangle. There are 27 icons on the screen, so that each icon has a shape of a 1×1 square. The *available* part of the screen is the part that does not contain any icon. What is the ratio of the area of the *available* part of the screen to the area of the screen?

(A) $\frac{1}{2}$ (B) $\frac{11}{20}$ (C) $\frac{13}{24}$ (D) $\frac{89}{143}$ (E) $\frac{2}{3}$

12. What is this sum equal to?

$$\frac{1}{2} + \frac{1}{3} + \frac{1}{6} + \frac{2}{3} \cdot \left(\frac{1}{2} + \frac{1}{3} + \frac{1}{6}\right) + \frac{1}{3} \cdot \left(\frac{1}{2} + \frac{1}{3} + \frac{1}{6}\right).$$

(A) 1   (B) 1.2   (C) 1.5   (D) 2   (E) 2.1

13. How many squares are there?

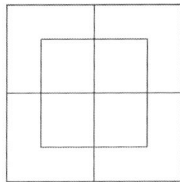

(A) 5   (B) 6   (C) 18   (D) 10   (E) 8

14. Little Luna has 15 red, 3 blue, and 2 orange unit cubes. She builds a tower from these cubes, placing each next cube on the previous one, so that the next cube is of a different color than the previous one. At most, how many cubes tall can Luna's tower be?

(A) 9   (B) 10   (C) 11   (D) 12   (E) 20

15. Given a paper square of side length 2 and a paper circle of radius 1. Which of the following shapes is not possible to form? Cutting shapes is not allowed.

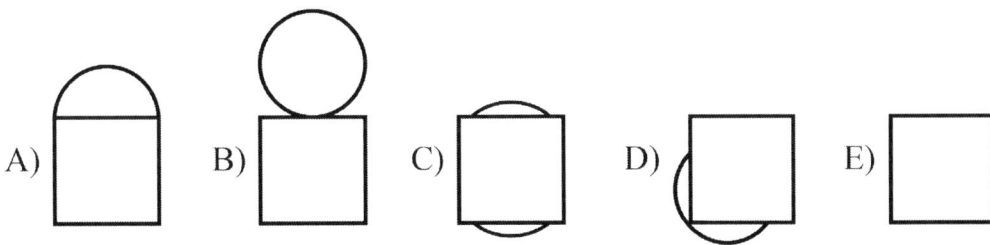

16. A big square is divided into four rectangles and one small square (see the picture). The perimeter of each rectangle is 500. What is the perimeter of the big square?

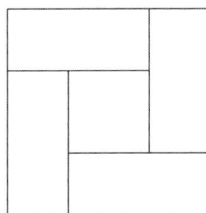

(A) 2024   (B) 1012   (C) 1000   (D) 888   (E) 1500

17. Given three shapes of areas a, b, c. Which of the following statements is true?

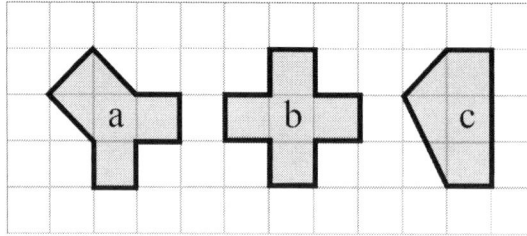

(A) $a = b$ (B) $b = c$ (C) $a > b$ (D) $a = c$ (E) $c > b$

18. Three bowls altogether contain 60 candies. The positive difference between the number of candies in any two bowls is either 4 or 8. Which of the following numbers corresponds to the number of candies in one of the bowls?

(A) 13   (B) 14   (C) 15   (D) 16   (E) 17

19. Five grams of paint is needed to color the surface of a cube of edge length 1 cm. Given a cube of edge length 4 cm and its eight corner cubes of edge length 1 cm are taken away. How many grams of paint is needed to color the surface of the leftover shape?

(A) 80   (B) 81   (C) 85   (D) 90   (E) 125

20. A kangaroo can jump either 3 meters or 5 meters. At least, how many jumps are needed to end up exactly 29 meters away from its starting place?

(A) 8   (B) 6   (C) 10   (D) 7   (E) 9

Part C: Each correct answer is worth 5 points

21. On Thursday Ann solved $\frac{1}{3}$ part of math homework exercises. On Friday she solved $\frac{1}{2}$ part of exercises left from Thursday. On Saturday she solved $\frac{1}{5}$ part of exercises left from Friday. On Sunday she solved all 4 exercises left from Saturday. How many problems did she solve on Thursday?

(A) 4   (B) 6   (C) 10   (D) 3   (E) 5

22. In the following Number Tower each number, except the bottom row numbers, is equal to the sum of two numbers directly below it. What is the value of $a + b$?

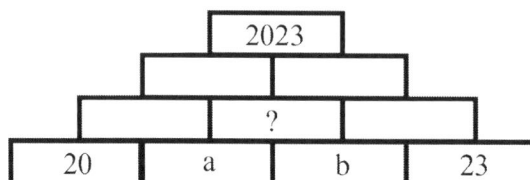

(A) 960   (B) 660   (C) 606   (D) 600   (E) 423

23. At least how many digits must a natural number have, so that erasing some of its digits we can get every possible two-digit number? For example, if the number is 4510 and we erase the digits 5 and 1, then we are left with the two-digit number 40.

(A) 19   (B) 17   (C) 15   (D) 14   (E) 12

24. In 2022, the grandfather had the same age as the last two digits of his birth-year, and the granddaughter had the same age as the last two digits of her birth-year. What was the sum of their ages in 2022?

(A) 70   (B) 72   (C) 75   (D) 76   (E) 80

25. The greatest and the smallest of the numbers $a + b$, $b + c$, $a + c$ are equal to 62 and 61, where a, b, c are natural numbers and they can be equal to each other. What is the sum of the digits of $a + b + c$?

(A) 8   (B) 9   (C) 10   (D) 11   (E) 12

26. Little Luna picked two different digits and kept them secret from her grandmother. Luna said the sum of these two digits to her grandmother, after that the grandmother said that she for sure knows both digits. What is the value of one of these two digits?

(A) 9   (B) 8   (C) 1 or 8   (D) 0   (E) 0 or 9

27. There are 28 domino tiles, so that the height of each tile is 0.4 cm and the width is 1 cm. We want to fit all tiles into a rectangular box, so that 28 tiles are placed in 4 layers and each layer includes 7 tiles. Tiles can be placed in each layer in both of the following ways:

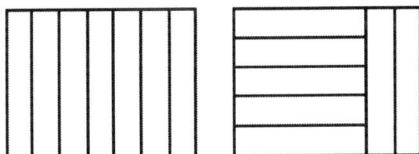

What is the smallest possible volume of such a box?

(A) 59 cm³   (B) 58 cm³   (C) 57 cm³   (D) 56 cm³   (E) 55 cm³

28. a and b are the remainders when natural numbers n and $4n$ are divided by 10 and 15, respectively. What is the greatest possible value of $a + b$?

(A) 19   (B) 20   (C) 21   (D) 22   (E) 23

29. Several pairs of natural numbers add up to 77. For example, 1 and 76, or 32 and 45. From all these pairs that add up to 77 we choose one pair of numbers that has the smallest possible LCM (least common multiple). What is the sum of the digits of their LCM?

(A) 10   (B) 11   (C) 12   (D) 13   (E) 15

30. The shaded rectangle is placed on a rectangle of side length 3 (see the figure). What is the ratio of the area of the rectangle of side length 3 to the area of the shaded rectangle?

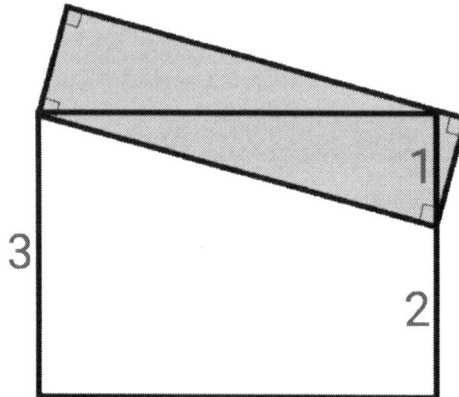

(A) 2:1   (B) 5:2   (C) 9:4   (D) 11:4   (E) 3:1

Test 7

Part A: Each correct answer is worth 3 points

1. At most how many days can there be in six consecutive months?

(A) 181   (B) 182   (C) 183   (D) 184   (E) 185

2. What is the coordinate of point A?

(A) 0.5   (B) 0.6   (C) 0.7   (D) 0.8   (E) 0.9

3. For the digit x we know that $3.0023 < 3.x023 < 3.6023$. What is the sum of all possible values of the digit x?

(A) 10   (B) 11   (C) 12   (D) 15   (E) 16

4. Math operations $\triangle$ and $\square$ are defined in the following way: $a \triangle b = a + b + 1$ and $a \square b = (a - 1) \cdot b$. What is $(1.1 \triangle 1.4) \square 2.6$?

(A) 6.5   (B) 4.5   (C) 4.3   (D) 4.2   (E) 3.6

5. Which picture shows $\angle BCA$?

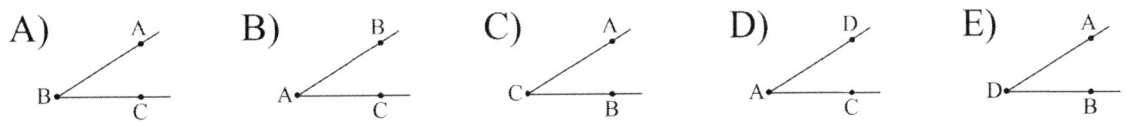

6. Some squares of a 6×6 square are numbered by numbers 1, 2,..., 18. The paper was folded across line l, then the paper was folded across line m. After this, with which square will the square number 4 coincide?

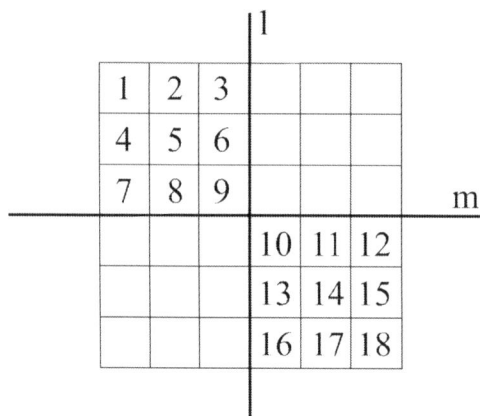

(A) 12   (B) 11   (C) 18   (D) 17   (E) 15

7. A local supermarket increases the price of its vanilla ice-cream from \$5 by 20 percent. At most, how many vanilla ice-creams can you buy with \$50?

(A) 10   (B) 9   (C) 8   (D) 7   (E) 6

8. A woodcutter needs 7.5 minutes to cut a 6 meters long wooden stick into 6 one meter long wooden sticks. How many minutes are needed to cut a 11 meters long wooden stick into 11 one meter long wooden sticks?

(A) 11.5   (B) 12.5   (C) 13   (D) 15   (E) 15.5

9. At least how many different digits must be used in order to write three consecutive three-digit numbers?

(A) 2   (B) 3   (C) 4   (D) 5   (E) 6

10. Given five different paper shapes (see the figure). Which two of these shapes can be put together to form a circle? Papers cannot be flipped.

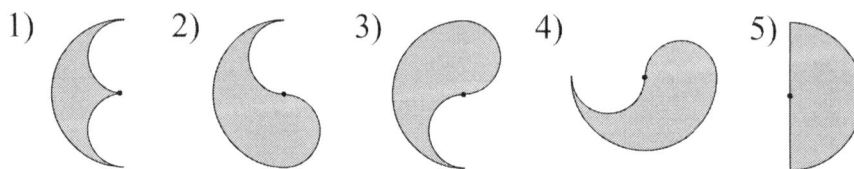

(A) 1 and 2   (B) 2 and 3   (C) 3 and 4   (D) 4 and 5   (E) 2 and 4

Part B: Each correct answer is worth 4 points

11. The sum of perimeters of triangle ABC and triangle PQR is equal to 20. What is the sum of perimeters of the shaded six triangles?

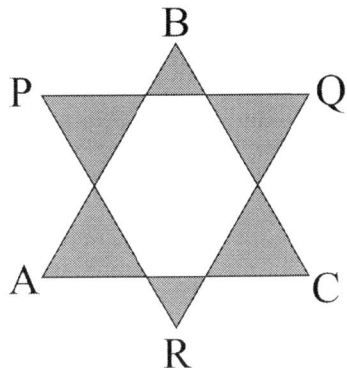

(A) 10   (B) 20   (C) 40   (D) 45   (E) 50

12. At most how many odd digits can the product of two two-digit numbers have?

(A) 0   (B) 1   (C) 2   (D) 3   (E) 4

13. Two cars started simultaneously toward each other from towns A and B, and they met in 1 hour. The car which came from A turned back and drove back to A, and the other car continued its way to A. Given that the car which came from A reached A 20 minutes earlier than the other car. How much time did the car which started from B spend on the whole trip from B to A?

(A) 1 hour 20 mins   (B) 2 hours 20 mins   (C) 2 hours 30 mins   (D) 3 hours   (E) 4 hours

14. In how many different ways is possible to write the numbers 1, 2, 3, 4 in the squares of a 2×2 square, so that the sum of the numbers of any two squares that share a side is a prime number?

(A) 8   (B) 10   (C) 16   (D) 20   (E) 24

15. A rectangle is divided into nine squares (see the figure), so that the side length of the smallest square is equal to 1. What is the side length of the greatest side of the rectangle?

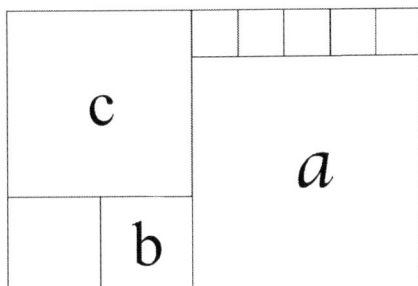

(A) 6   (B) 7   (C) 8   (D) 9   (E) 10

16. In each square of a 3×3 is written a number (see the figure). The places of two numbers were switched, so that the product of all numbers of the shaded squares is equal to the product of all numbers of the white squares. What is the sum of these two numbers?

6	5	1
2	7	10
14	9	12

(A) 12   (B) 17   (C) 19   (D) 21   (E) 24

17. A toy kangaroo is on the bottom leftmost square of a chessboard. It can only jump horizontally or vertically either 3 squares or 2 squares. At least, how many jumps are needed to get to the top rightmost square?

(A) 3   (B) 4   (C) 5   (D) 6   (E) 7

18. A rectangle of whole number side lengths is divided into two rectangles of whole number side lengths (see the figure), so that the perimeters of the small rectangles are 20 and 24. What is the smallest possible perimeter of the initial rectangle?

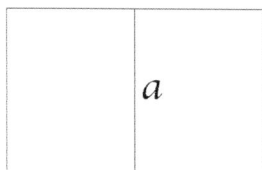

a

(A) 20   (B) 22   (C) 24   (D) 26   (E) 44

19. How many rectangles are there? (see the figure, a square is also a rectangle).

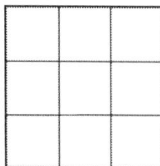

(A) 27   (B) 30   (C) 32   (D) 35   (E) 36

20. From the list of the numbers 1, 2,..., 100 all numbers divisible by 2 are erased, then all numbers divisible by 3 are erased, then all numbers divisible by 4 are erased (if there are any left), and so on. What is the product of the digits of the last erased number?

(A) 10   (B) 20   (C) 30   (D) 62   (E) 63

Part C: Each correct answer is worth 5 points

21. In how many different ways is possible to cut (by a straight line) a 4×6 paper rectangle into two parts, so that when one part is folded it becomes the same as the other part?

(A) 0   (B) 1   (C) 2   (D) 3   (E) 4

22. What change in percent is made to the area of a rectangle by decreasing its length and its perimeter by 10 percent?

(A) 25   (B) 26   (C) 20   (D) 19   (E) 21

23. Two cyclists start simultaneously toward each other from cities A and B. One going at 14 miles per hour. Given that when they were 10 miles apart, 40 minutes after that they were again 10 miles apart. What is the speed (in miles per hour) of the other cyclist?

(A) 12   (B) 14   (C) 15   (D) 16   (E) 17

24. At least how many numbers must be added to the list of the numbers 15, 20, 80, so that the positive difference of any two of the numbers belongs to the new list of these numbers?

(A) 1   (B) 5   (C) 10   (D) 12   (E) 13

25. A three-digit number is called "beautiful" if the difference of that three-digit number and the product of its digits is equal to 110. How many "beautiful" three-digit numbers are there?

(A) 8   (B) 10   (C) 9   (D) 12   (E) 11

26. In how many different ways is possible to divide the numbers 1, 2, 3, 4, 5, 6, 7, 8 into four pairs, so that the sum of the numbers in any pair is a prime number?

(A) 6   (B) 7   (C) 8   (D) 9   (E) 10

27. a, b, c, d is a rearrangement of the numbers 1, 2, 3, 4, so that:
a is divisible by b,
$a + b$ is divisible by c,
$a + b + c$ is divisible by d.
How many such rearrangements are there?

(A) 0   (B) 1   (C) 2   (D) 3   (E) 4

28. At least how many digits must a natural number have, so that erasing some of its digits we can get every possible two-digit number which is divisible by 3? For example, if the number is 9239 and we erase the digits 2 and 3, then we are left with the two-digit number 99.

(A) 15   (B) 16   (C) 17   (D) 18   (E) 19

29. a, b, c, d are different positive digits. Eleven two-digit numbers were chosen from the following twelve two-digit numbers $\overline{ab}$, $\overline{ac}$, $\overline{ad}$, $\overline{bc}$, $\overline{ba}$, $\overline{bd}$, $\overline{ca}$, $\overline{cb}$, $\overline{cd}$, $\overline{da}$, $\overline{db}$, $\overline{dc}$, so that the sum of chosen numbers is equal to 450. What is the sum of the digits of the two-digit number that was not chosen?

(A) 3   (B) 4   (C) 5   (D) 6   (E) 7

30. The numbers 1, 2, 3, 4, 5, 6 are written on the faces of a cube (one number per face). For any two faces that share a common edge the positive difference of the numbers written on these two faces is written on that edge. What is the greatest possible sum of all numbers written on all 12 edges?

(A) 30   (B) 31   (C) 32   (D) 33   (E) 34

Answers

Problem	Practice test 1	Practice test 2	Practice test 3	Practice test 4
1	(A)	(C)	(C)	(C)
2	(E)	(D)	(C)	(D)
3	(B)	(C)	(B)	(D)
4	(D)	(B)	(D)	(D)
5	(B)	(D)	(C)	(C)
6	(E)	(E)	(C)	(D)
7	(A)	(D)	(D)	(D)
8	(D)	(C)	(B)	(A)
9	(E)	(B)	(A)	(B)
10	(D)	(A)	(E)	(B)
11	(D)	(E)	(E)	(C)
12	(C)	(D)	(E)	(A)
13	(D)	(D)	(D)	(B)
14	(E)	(C)	(D)	(D)
15	(D)	(A)	(B)	(D)
16	(A)	(A)	(D)	(E)
17	(C)	(C)	(D)	(D)
18	(E)	(E)	(B)	(A)
19	(C)	(C)	(D)	(A)
20	(C)	(D)	(D)	(E)
21	(E)	(B)	(D)	(E)
22	(C)	(C)	(A)	(D)
23	(D)	(B)	(D)	(B)
24	(C)	(B)	(D)	(B)
25	(D)	(C)	(E)	(D)
26	(B)	(A)	(B)	(C)
27	(C)	(E)	(A)	(A)
28	(B)	(D)	(C)	(C)
29	(C)	(B)	(B)	(B)
30	(D)	(D)	(D)	(C)

Problem	Practice test 5	Practice test 6	Practice test 7
1	(C)	(C)	(D)
2	(D)	(B)	(C)
3	(B)	(E)	(D)
4	(D)	(C)	(A)
5	(C)	(E)	(C)
6	(B)	(C)	(E)
7	(C)	(C)	(C)
8	(B)	(E)	(D)
9	(A)	(E)	(B)
10	(D)	(B)	(E)
11	(E)	(D)	(B)
12	(E)	(D)	(E)
13	(C)	(D)	(B)
14	(B)	(C)	(A)
15	(E)	(C)	(D)
16	(B)	(C)	(B)
17	(B)	(D)	(D)
18	(D)	(D)	(D)
19	(D)	(A)	(E)
20	(C)	(E)	(E)
21	(C)	(E)	(C)
22	(B)	(B)	(D)
23	(B)	(A)	(D)
24	(E)	(B)	(E)
25	(B)	(D)	(B)
26	(C)	(E)	(A)
27	(E)	(D)	(C)
28	(D)	(B)	(B)
29	(D)	(C)	(A)
30	(C)	(E)	(C)

Solutions of Test 1

Part A: Each correct answer is worth 3 points

1. Each row and each column of a 4×4 square consists of these four shapes ●, △, □, ○. In which of the squares a, b, c, d, e is the shape ○?

(A) a (B) b (C) c (D) d (E) e

Answer. (A)

Solution. Here is the correct order where each shape must be (see the figure).

The correct answer is (A).

---

2. There are four pieces of paper on the table. What is their correct order (from bottom to top)?

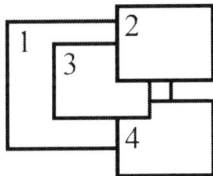

(A) 1, 2, 3, 4   (B) 2, 1, 3, 4   (C) 3, 2, 4, 1   (D) 4, 3, 2, 1   (E) 1, 4, 3, 2

Answer. (E)

Solution. We have

4 is above 1,

3 is above 4,

2 is above 3.

So, their correct or from bottom to top is 1, 4, 3, 2. The correct answer is (E).

3. The sum of the number of floors that are below and above Luna's apartment is 13. How many floors are there in Luna's building?

(A) 13   (B) 14   (C) 15   (D) 16   (E) 20

Answer. (B)

Solution. The number of all floors is equal to 13 plus the floor where Luna lives, so $13 + 1$. The correct answer is (B).

4. Two transparent paper squares are colored like this.

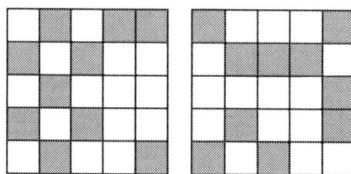

Placing these squares on each other, which of the following squares is not possible to get?

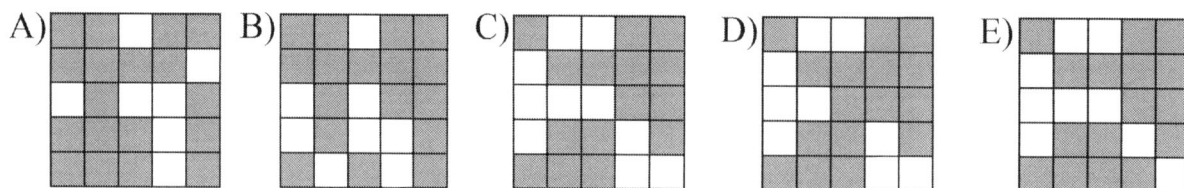

Answer. (D)

Solution. The central squares of both paper squares are not black, while the central square of square (D) is black. So, square (D) is not possible to get by placing these squares on each other. One can easily check that any of the squares (A), (B), (C), (E) is possible to get. The correct answer is (D).

5. In two places in between of the digits of the number 20245 plus (+) signs were placed, so that the sum is the smallest possible (out of all possible such sums). What is that sum?

(A) 43   (B) 49   (C) 67   (D) 211   (E) 247

> Answer. (B)
> **Solution.** List all possible sums.
> $2 + 0 + 245 = 247$.
> $20 + 2 + 45 = 67$.
> $20 + 24 + 5 = 49$.
> $202 + 4 + 5 = 211$.
> So, the smallest possible sum (out of all possible sums) is 49. The correct answer is (B).

6. How many digits are there, so that you can move one stick to get another digit?

$$0123456789$$

(A) 2   (B) 3   (C) 4   (D) 5   (E) 6

> Answer. (E)
> **Solution.** One can easily check that 1, 4, 7, 8 do not work. 0, 2, 3, 5, 6, 9 work (see the figure).
>
> $$6 \leftrightarrow 0 \leftrightarrow 9, 2 \leftrightarrow 3 \leftrightarrow 5$$
>
> So, there are 6 such digits. The correct answer is (E).

7. There are dozens of pens in the box. You are allowed to choose some of these pens and divide the chosen pens into two groups, so that the number of pens in each group is a prime number. Then, you need to move one or more pens from the first group to the second group, so that after this move the number of pens in each group is a prime number. What is the smallest number of pens that you can move so that all these conditions are true?

(A) 1   (B) 2   (C) 3   (D) 4   (E) 10

> Answer. (A)
> **Solution.** Let p and q be the initial number of pens in each group, where p and q are prime numbers. You need to move one or more pens, so if you can move 1 pen so that all the conditions of the problems are true, then the smallest such number is 1 and this ends the solution. After moving 1 pen from the first group to the second there will be $p - 1$ pens in the first group and $q + 1$ pens in the second group, where $p - 1$ and $q + 1$ must be prime numbers.
> p is a prime number and $p - 1$ is a prime number, so p must be 3.
> q is a prime number and $q + 1$ is a prime number, so q must be 2.
> Then, the smallest number of pens that you can move so that all the conditions are true is 1.
> So, you need to choose 3 pens and 2 pens (both are prime numbers), then move 1 pen from the first group to the second group to have 2 pens and 3 pens. The correct answer is (A).

8. Which path from A to B is the shortest?

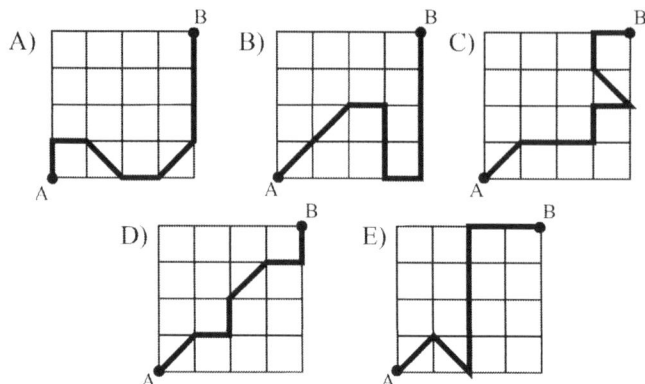

Answer. (D)

Solution. All paths include two diagonals and several sides. The smallest number of sides has (D), only 4 sides. The correct answer is (D).

9. Which of these shapes is not possible to divide into two such shapes?

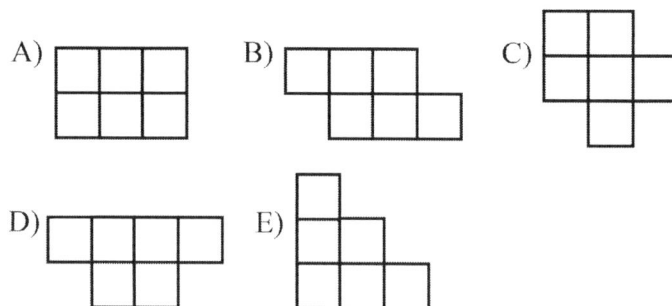

Answer. (E)

Solution. (E) is not possible, because the shaded square is possible to cover by such only in one way (see the figure).

(A), (B), (C), (D) are possible (see the figure). The correct answer is (E).

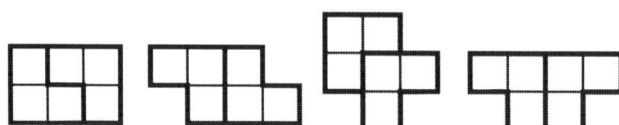

56

10. What is the sum of the digits of the smallest natural number that is not divisible by any of its digits?

(A) 9   (B) 7   (C) 6   (D) 5   (E) 4

> **Answer. (D)**
> **Solution.** Any one-digit number is divisible by "its digit". Any of the numbers 11, 12,..., 19 is divisible by one of its digits, namely by the digit 1. Any of the numbers 20, 21, 22 is divisible by one of its digits, namely 20 and 22 are divisible by 2, and 21 is divisible by 1. So, 23 is the smallest natural number that is not divisible by any of its digits, and the sum of its digits is $2 + 3 = 5$. The correct answer is (D).

Part B: Each correct answer is worth 4 points

11. Phil is a darts player, at each turn he gets 10 out of 10, or 9 out of 10. Phil played 10 turns and he did not get the same score at any three turns in a row. What is Phil's smallest possible score after these 10 turns?

(A) 90   (B) 91   (C) 92   (D) 93   (E) 94

> **Answer. (D)**
> **Solution.** In at least one of his first three turns (1-3) he got a 10, in at least one of the next three turns (4-6) he got a 10, in at least one of the next three turns (7-9) he got a 10. So, altogether he got at least three 10 and seven 9, this means that his overall score is at least $3 \cdot 10 + 7 \cdot 9 = 93$. An example so that he scored 93 and the condition of the problem is true (he did not get the same score at any three turns in a row): 9, 9, 10, 9, 9, 10, 9, 9, 10, 9. The correct answer is (D).

12. Luna's apartment is in one of the floors of a multi-storey building. The sum of the numbers of all other floors (except her floor) is 17. How many floors are there in the building?

(A) 4   (B) 5   (C) 6   (D) 7   (E) 8

> **Answer. (C)**
> **Solution.** This multi-storey building cannot have 7 or more floors, because $1+2+3+4+5+6 = 21 > 17$. It cannot have 5 or less floors, because $2+3+4+5 = 14 < 17$. So, this multi-storey building has 6 floors and Luna's apartment is in $1+2+3+4+5+6-17 = 4$ floor. The correct answer is (C).

13. A whole number n is equal to the sum of two two-digit multiples of 5. What is the greatest possible value of the sum of the digits of n?

(A) 10   (B) 12   (C) 13   (D)   (E)

Answer. (D)
Solution. Consider the following two cases.
Case 1. If the sum of the unit digits of these two two-digit numbers is not more than 5. Then, the sum of the digits of n is not more than $9 + 5 = 14$.
Case 2. If the sum of the unit digits of these two two-digit numbers is 10, then the sum of the digits of n is not more than $10 + 0 = 10$.
For example, for 95 we have $9 + 5 = 14$ and $95 = 35 + 60$. So, the greatest possible value of the sum of the digits of n is 14 (if we take $n = 95$). The correct answer is (D).

14. What is the area of the shaded shape? The area of each square is 1.

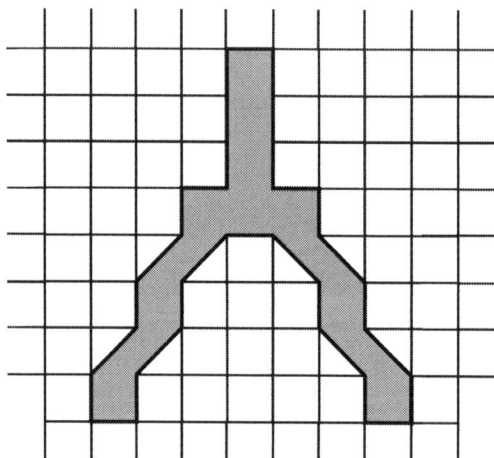

(A) 11.5   (B) 12   (C) 12.5   (D) 13   (E) 14

Answer. (E)
Solution. The shaded shape consists of 10 squares and 8 triangles. Two triangles together form one square, so the area of shaded shape is equal to the area of $10 + 4$ squares. The correct answer is (E).

15. Three-sided football is a variation of football played with three teams instead of the usual two (see the figure). After each game only one team wins and the other two lose. 101 teams took part in a three-sided football tournament, where the losing team leaves. One of the teams won the tournament, how many games were played?

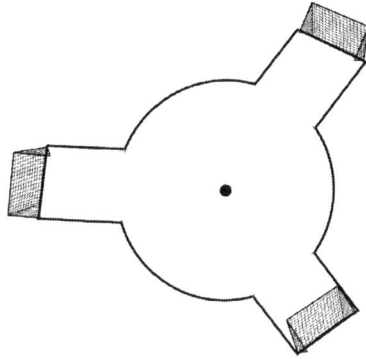

(A) 33   (B) 34   (C) 35   (D) 50   (E) 51

Answer. (D)
Solution. After each game two teams lose, so after each game two teams leave. So, the number of games played is $\frac{101-1}{2} = 50$. The correct answer is (D).

16. A three-digit number is 11 times greater than the sum of its digits. What is the product of its digits?

(A) 72   (B) 60   (C) 45   (D) 30   (E) 18

Answer. (A)
Solution. Let $\overline{abc}$ be that three-digit number. Given that

$$\overline{abc} = 11 \cdot (a + b + c).$$

We get

$$100 \cdot a + 10 \cdot b + c = 11 \cdot a + 11 \cdot b + 11 \cdot c.$$

Then

$$89 \cdot a = 10 \cdot c + b = \overline{cb}.$$

As $\overline{cb}$ is a two-digit, then from last equation we get $a = 1$ and $\overline{cb} = 89$. So $a \cdot b \cdot c = 1 \cdot 9 \cdot 8 = 72$. The correct answer is (A).

17. Using only one pair of parentheses in the expression $4 \cdot 2 + 20 : 4$ we can get different expressions. What is the greatest possible value among these expressions?

(A) 7   (B) 22   (C) 28   (D) 30   (E) 32

Answer. (C)

Solution. Using only one pair of parentheses in the expression $4 \cdot 2 + 20 : 4$ we can get the following expressions:

$$(4 \cdot 2 + 20) : 4 = 7,$$

$$4 \cdot (2 + 20 : 4) = 28,$$

$$4 \cdot (2 + 20) : 4 = 22.$$

So, the greatest possible value among these expressions is 28. The correct answer is (C).

18. A, B, C, D are points on a plane, so that $AB = 3$, $BC = 30$, $AC = 33$, $AD = 23$, $BD = 20$. What is the length CD?

(A) 3   (B) 6   (C) 8   (D) 9   (E) 10

Answer. (E)

Solution. Given that $AB = 3$, $BC = 30$, $AC = 33$, so $AB + BC = AC$. This means that point B belongs to line segment AB. Given also that $AB = 3$, $BD = 20$, $AD = 23$, so $AB + BD = AD$. This means that point B belongs to line segment AD. As $AD < AC$, then point D belongs to line segment AC. We get $AD + DC = AC$, then $CD = 33 - 23 = 10$. The correct answer is (E).

19. All these shapes consist of 12 cubes. Which of these shapes is not possible to divide into four such shapes?

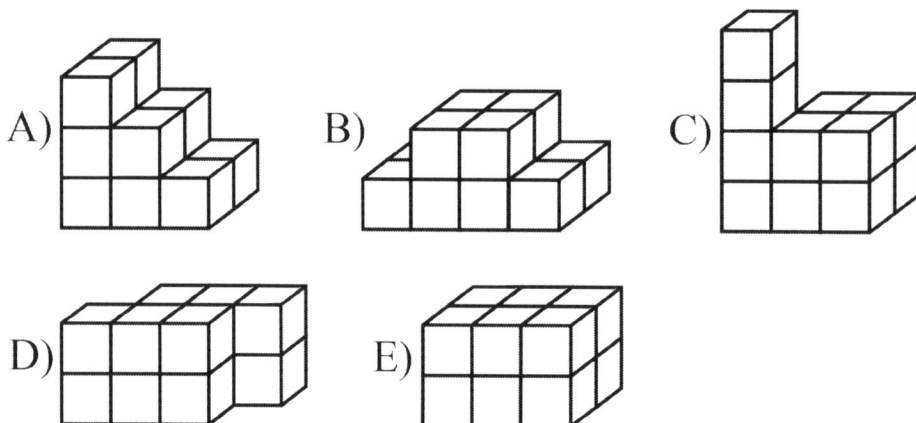

60

Answer. (C)

Solution. (A), (B), (D), (E) are possible.

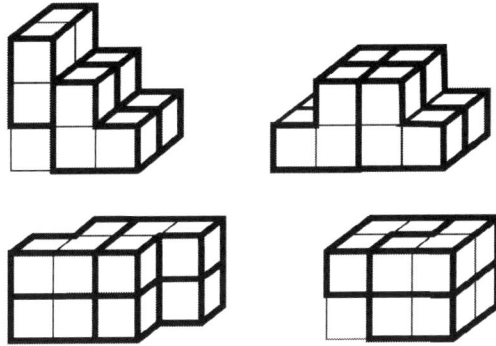

(C) is not possible, because its top two cubes cannot be a part of such .

The correct answer is (C).

20. A little boy sitting next to the river noticed that two groups of ducks are swimming in the river. In the first group 13 ducks, in the second group 7 ducks. From time to time some ducks were moving from one group to the other one. After some time the number of ducks in both groups was the same. What is the greatest possible number of ducks, so that each of them made an odd number of moves from one group to another?

(A) 3   (B) 10   (C) 17   (D) 18   (E) 20

Answer. (C)

Solution. If a duck made an even number of moves from one group to another, then we assume that the number of its moves is 0. If a duck made an odd number of moves from one group to another, then we assume that the number of its moves is 1. The greatest possible number of ducks that moved from the first group is 10, the greatest possible number of ducks that moved from the second group is 7. Then, the greatest possible number of ducks, so that each of them made an odd number of moves from one group to another is $10 + 7 = 17$.

The correct answer is (C).

Part C: Each correct answer is worth 5 points

21. A scale is broken and gives a reading that is 10 percent less than the actual weight of the item. It shows that a bag of pears weights 5.4 pounds. What is the actual weight (in pounds) of this bag of pears?

(A) 4.86   (B) 5   (C) 5.1   (D) 5.15   (E) 6

Answer. (E)
Solution. A scale is broken and gives a reading that is 10 percent less than the actual weight of the item, this means that 90 percent of the actual weight of this bag of pears is 5.4 pounds. So, 10 percent of the actual weight of this bag of pears is $\frac{5.4}{9} = 0.6$ pounds. Then, 100 percent of the actual weight of this bag og pears is $5.4 + 0.6 = 6$ pounds. The correct answer is (E).

22. A girl standing in a garden is 5 meters apart from a bird and she is 4 meters apart from a rabbit. At least how far apart can the bird and the rabbit be?

(A) 0.5 meters   (B) 0.8 meters   (C) 1 meter   (D) 2 meters   (E) 9 meters

Answer. (C)
Solution. As the girl is 5 meters apart from a bird and is 4 meters apart from a rabbit, then the smallest distance between the bird and the rabbit is possible only in the following case (see the figure). So, the bird and the rabbit are at least 1 meter apart. The correct answer is (C).

23. The sum of two different positive whole numbers is six times their greatest common factor. What is the largest possible quotient when the largest of these numbers is divided by the smallest of these numbers?

(A) 2   (B) 3   (C) 4   (D) 5   (E) 10

Answer. (D)
Solution. Let d be the greatest common factor of these two different positive whole numbers. So, let these numbers be $d \cdot m$ and $d \cdot n$, where m and n are different positive whole numbers. Assume that $m > n$. Given that their sum is six times their greatest common factor, so

$$d \cdot m + d \cdot n = 6 \cdot d.$$

Then, we get $m + n = 6$. As we want to find the largest possible quotient when the largest of these numbers is divided by the smallest of these numbers, then $m = 5$ and $n = 1$. So, we get $\frac{5 \cdot d}{d} = 5$. The correct answer is (D).

24. A river fence consists of six vertical columns and some steel pickets. There are 9 steel pickets in part A, 12 steel pickets in part E. The difference of the number of pickets in any two neighboring parts is not more than 2. What is the greatest possible number of all pickets in parts B, C, D altogether?

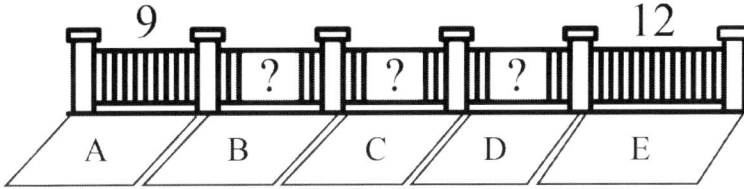

(A) 33   (B) 37   (C) 38   (D) 39   (E) 40

Answer. (C)
Solution. Given that there are 9 steel pickets in part A, 12 steel pickets in part E, and the difference of the number of pickets in any two neighboring parts is not more than 2, then the number of pickets in part B is not more than 11, the number of pickets in part C is not more than 13, and the number of pickets in part D is not more than 15. Moreover, in part D there cannot be 15. So, the greatest possible number of all pickets in parts B, C, D altogether is at most $11 + 13 + 14 = 38$. Note that 38 is possible, for example 9, 11, 13, 14, 12. The correct answer is (C).

25. Ann has two dimes and three quarters. Bob has two dimes. What is the sum of all possible amounts (in cents) that Ann can pay to Bob?

(A) 950   (B) 860   (C) 855   (D) 780   (E) 740

Answer. (D)
Solution. Altogether Ann has 95 cents $(10 + 10 + 25 + 25 + 25)$. So, after paying any amount to Bob she cannot end up with 5 cents or 15 cents. This means that she cannot pay 90 cents or 80 cents. So, the sum of all possible amounts that Ann can pay to Bob is $5 + 10 + ... + 95 - 90 - 80 = 780$. The correct answer is (D).

26. Given $\angle ABC = 120°$. Each of rays BD and BE divide angle ABC into two angles, so that the ratio of their measures is $1 : 5$ (see the figure). What is the measure of $\angle DBE$?

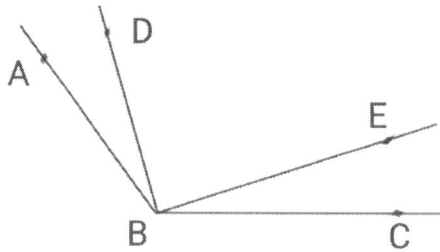

(A) 90°   (B) 80°   (C) 70°   (D) 65°   (E) 60°

Answer. (B)

Solution. As the ratio of the measure of $\angle ABD$ to the measure of $\angle DBC$ is 1:5, then the measure of $\angle ABD$ is $\frac{1}{6}$ part of the measure of $\angle ABC$. In a similar way, the measure of $\angle CBE$ is $\frac{1}{6}$ part of the measure of $\angle ABC$. So, the measure of $\angle DBE$ is $1 - \frac{1}{6} - \frac{1}{6} = \frac{2}{3}$ part of the measure of $\angle ABC$. As $\angle ABC = 120°$, then we get $120° \cdot \frac{2}{3} = 80°$. The correct answer (B).

27. Annie has several suitcases. There are two suitcases in one of them, there is one suitcase in two of them, and there is no suitcase in the remaining two suitcases. How many suitcases does Annie have?

(A) 3   (B) 4   (C) 5   (D) 6   (E) 7

Answer. (C)

Solution. As there is no suitcase in two of her suitcases, then all suitcases must be placed in two suitcases (see the figure).

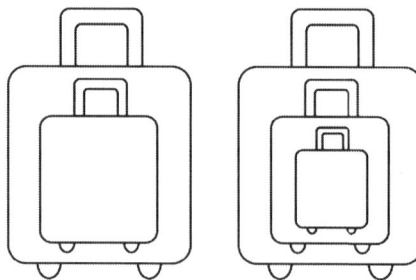

There is one suitcase in one of these suitcases and there are two suitcases in the other one. So, Annie has 5 suitcases. The correct answer is (C).

Alternative solution. The suitcases are divided into three groups, 1, 2, 2. So, there are $1 + 2 + 2 = 5$ suitcases. The correct answer is (C).

28. River current speed is 2.4 meters per second. Moving with the river current ducks cross the bridge in 2 seconds, and moving against the river current ducks cross the bridge in 5 seconds. How wide is the bridge?

(A) 8 meters   (B) 16 meters   (C) 18 meters   (D) 14 meters   (E) 12 meters

Answer. (B)

Solution. Let the speed of ducks on a still water be x meters per second. Given that moving with the river current ducks cross the bridge in 2 seconds and as the river current speed is 2.4 meters per second, then the width of the bridge is $2 \cdot (x+2.4)$ meters. Given that moving against the river current ducks cross the bridge in 5 seconds and the river current speed is 2.4 meters per second, then the width of the bridge is $5 \cdot (x-2.4)$ meters. We get $2 \cdot (x+2.4) = 5 \cdot (x-2.4)$. We get $3x = 7 \cdot 2.4$. So $x = 5.6$ meters per second and $2 \cdot (x + 2.4) = 16$ meters. The correct answer is (B).

29. In a shooting tournament Mia shot seven times and each time she got one of the following scores 0, 1, 2, ..., 10. Given that the sum of the scores of any three consecutive shots are different from each other. What is the greatest possible overall score that Mia could get after her seven shots?

(A) 64   (B) 65   (C) 66   (D) 67   (E) 68

Answer. (C)

Solution. As there were seven shots, then altogether there are five such sums (of three consecutive shots). Given that all these five sums are different, then among these five sums there is a sum that is not more than 26. So, the overall score that Mia could get after her seven shots is not more than 66. She could get 66, for example, in the following way 10, 10, 10, 9, 9, 8, 10. The correct answer is (C).

30. At most how many squares of a 4×4 square can be painted, so that the painted squares do not form any such shape ?

(A) 8   (B) 9   (C) 10   (D) 11   (E) 12

Answer. (D)

Solution. We provide an example where 11 squares are painted, so that the painted squares do not form any such shape.

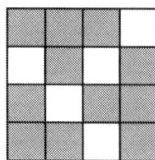

Now, we prove that if at least 12 squares are painted, then there is at least one such form. If all squares of square A are painted, then it is obvious that there is at least one such form.

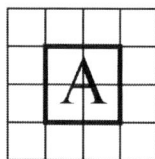

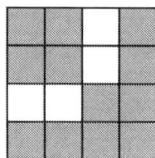

If at least one of the squares of square A is not painted (see the figure),

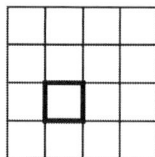

then divide this 4×4 square to four 2×2 squares. If all four squares of one of these four 2×2 squares are painted, then it is obvious that there is at least one such form.

If for each of these four 2×2 squares three squares are painted, then we get this case.

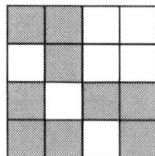

Note that no matter how we paint three square of a blank 2×2 square, we get at least one such shape. This ends the proof. The correct answer is (D).

Solutions of Test 2

Part A: Each correct answer is worth 3 points

1. The international Morse code encodes the 26 letters of the English alphabet and all nine digits from 0 to 9 (see the figure).

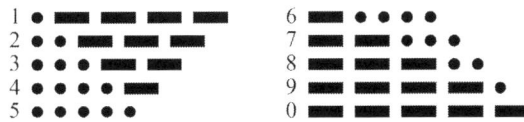

What is the value of this expression?

(A) 15   (B) 16   (C) 17   (D) 18   (E) 19

> Answer. (C)
> **Solution.** The figure shows that using the international Morse code the given expression can be decoded as $2 + 6 + 9$. So, we get $2 + 6 + 9 = 17$. The correct answer is (C).

2. A garden is in the shape of a rectangle. 50 trees are planted 2 meters apart along its perimeter, so that there is one tree in every corner of the garden. What is its perimeter?

(A) 43 meters   (B) 50 meters   (C) 98 meters   (D) 100 meters   (E) 150 meters

> Answer. (D)
> **Solution.** Given that altogether there are 50 trees along the perimeter of the rectangle and trees are planted 2 meters apart, so the perimeter of the rectangle is $50 \cdot 2 = 100$ meters. The correct answer is (D).

67

3. Little Luna has built a tower using 7 green cubes and a few white cubes, so that the number of white cubes is less than 7. She did not put two cubes of the same color on each other. How many cubes are there?

(A) 8   (B) 10   (C) 13   (D) 14   (E) 15

Answer. (C)
Solution. Given that little Luna did not put two cubes of the same color on each other. For simplicity let W be a white cube and G be a green cube. If she started with a white cube, then she could get either this tower

$$W, G, W, G, W, G, W, G, W, G, W, G, W, G,$$

or this tower

$$W, G, W, G, W, G, W, G, W, G, W, G, W, G, W.$$

In both cases, the number of white cubes is not less than 7. So, this means that Luna did not start with a white cube. If she started with a green cube, then she could get either this tower

$$G, W, G, W, G, W, G, W, G, W, G, W, G,$$

or this tower

$$G, W, G, W, G, W, G, W, G, W, G, W, G, W.$$

In the last case the number of white cubes is not less than 7. So, only this case is possible $G, W, G, W, G, W, G, W, G, W, G, W, G$. Then, we get that there are 7 green cubes and 6 white cubes. So, the number of cubes is $7 + 6 = 13$. The correct answer is (C).

4. Fill in each circle with an even digit, so that all inequalities are correct (see the figure).

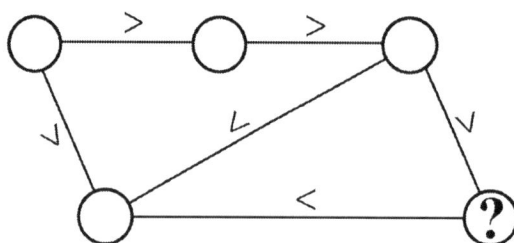

What digit must be in the rightmost circle?

(A) 0   (B) 2   (C) 4   (D) 6   (E) 8

Answer. (B)

Solution. Note that 8 (the largest even digit) and 0 (the smallest even digit) must be in the following circles (see the figure).

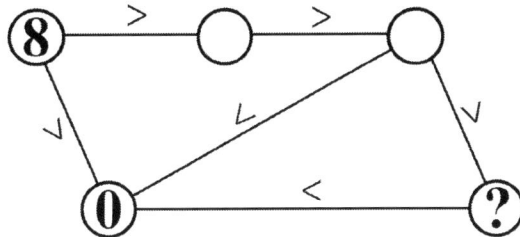

Note also that 6, 4, and 2 must be in the following circles (see the figure).

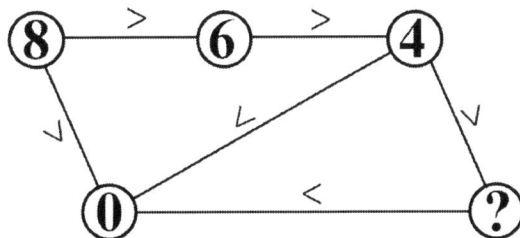

So, in the rightmost circle must be the digit 2. The correct answer is (B).

5. For natural number n we know that

$$\frac{4}{5} + \frac{2}{7} = \frac{n}{70}.$$

What is n?

(A) 67 (B) 70 (C) 72 (D) 76 (E) 78

Answer. (D)

Solution. We have

$$\frac{4}{5} + \frac{2}{7} = \frac{4 \cdot 7 + 2 \cdot 5}{35} = \frac{38}{35} = \frac{76}{70}.$$

Given also that

$$\frac{4}{5} + \frac{2}{7} = \frac{n}{70}.$$

From the last two equations we get

$$\frac{76}{70} = \frac{n}{70}.$$

This means that $n = 70$. The correct answer is (D).

6. How many digits are there, so that you can move one stick to get another digit?

$$0123456789$$

(A) 0   (B) 1   (C) 2   (D) 3   (E) 4

Answer. (E)
Solution. We do not get another digit by moving one stick from any of the digits 0, 1, 2, 3, 4, 5. For the digits 6, 7, 8, 9 to get another digit one stick can be moved like this:

$$6 \rightarrow 5, \ 7 \rightarrow 1, \ 8 \rightarrow 0, \ 9 \rightarrow 3$$

7. How many of these shapes can be the common part of two triangles?

(A) 0   (B) 1   (C) 2   (D) 3   (E) 4

Answer. (D)
Solution. If two triangles have a common part, then the common part is either a point or must have a border that is a straight line. So, the common part cannot be a circle. The common part can be three other shapes, for example:

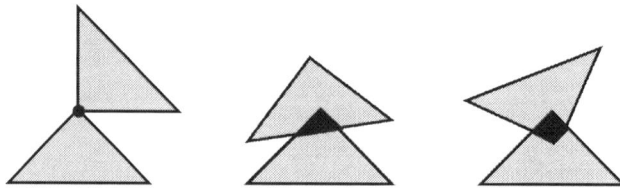

So, 3 of these shapes can be the common part of two triangles. The correct answer is (D).

8. Annie has three nickels (5 cent coin), three dimes (10 cent coin), and three quarters (25 cent coin). At most, how many coins can she give to pay a bill of 90 cents?

(A) 5   (B) 6   (C) 7   (D) 8   (E) 9

> Answer. (C)
>
> **Solution.** Note that Annie needs to use at least two quarters (25 cent coin), because if she does not use at least two quarters than
>
> $$3 \cdot 5 + 3 \cdot 10 + 25 = 70 < 90.$$
>
> As Annie used at least two quarters, let us consider the following two cases.
> If Annie used three quarters, then the number of used coins is not more than six.
> If Annie used two quarters, then the number of used coins is not more than seven. Note that $90 = 2 \cdot 5 + 3 \cdot 10 + 2 \cdot 25$. So, Annie can at most give seven coins to pay a bill of 90 cents. The correct answer is (C).

9. A clown has green, red, and yellow balloons. Exactly six of them are not red and exactly four of them are not yellow. How many more yellow balloons are there than red balloons?

(A) 1   (B) 2   (C) 3   (D) 4   (E) 5

> Answer. (B)
>
> **Solution.** Given that exactly six of them are not red, so there are altogether six green and yellow balloons. Given also that exactly four of them are not yellow, so there are altogether four green and red balloons. This means that, there the number of yellow balloons is two more than the number of red balloons. The correct answer is (B).

10. Given a square of side length 4. What is the area of the shaded part? (see the figure).

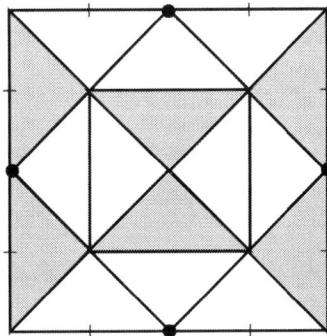

(A) 6   (B) 5   (C) 4   (D) 3   (E) 2.8

Answer. (A)

Solution. This square can be divided into 16 congruent triangles (see the figure).

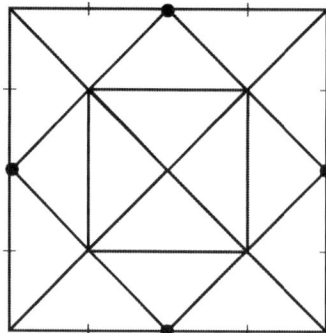

As altogether there are 16 triangles and 6 of them must be shaded, then the area of the shaded part is equal to $\frac{6}{16}$ part of square's area, that is

$$\frac{6}{16} \cdot 4 \cdot 4 = 6.$$

So, the area of the shaded part is equal to 6. The correct answer is (A).

Part B: Each correct answer is worth 4 points

11. Mira drew three triangles (her triangles can have a common side). Which of the following numbers cannot be equal to the number of different sides of these triangles?

(A) 9   (B) 8   (C) 7   (D) 6   (E) 4

Answer. (E)

Solution. (A), (B), (C), (D) are possible (see the figure).

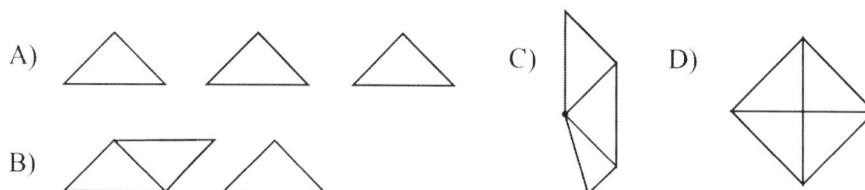

(E) is not possible, because one triangle has three sides. If we draw a line segment which is different than these three sides, then we cannot get another triangle by these four line segments. The correct answer is (E).

72

12. In how many different ways is it possible to create three pairs from the numbers 1, 2, 3, 4, 5, 6, so that the difference of the largest and the smallest numbers of each pair is a prime number?

(A) 0   (B) 1   (C) 2   (D) 3   (E) 4

> Answer. (D)
> **Solution.** The numbers 6 and 2 must be in different pairs, because $6 - 2 = 4$ and 4 is not a prime number. Consider the following two cases.
> **Case 1.** If the numbers 1 and 6 are in the same pair, then 3 and 5 are in the same pair, and 2 and 4 are in the same pair. So, we get $(1, 6), (3, 5), (2, 4)$.
> **Case 2.** If the numbers 1, 2, 6 are in different pairs, then 5 and 2 are in the same pair. In this case, the following two options are possible $(2,5), (1,4), (6,3)$ and $(2, 5), (1, 3), (6, 4)$.
> So, overall there are three different ways to create such pairs. The correct answer is (D).

13. Given two positive whole numbers, so that one of them is divisible by the other one. We know that the dividend is five times more than the divisor, and that the divisor is twice more than the quotient. What is the sum of the dividend, divisor, and quotient?

(A) 50   (B) 55   (C) 60   (D) 65   (E) 70

> Answer. (D)
> **Solution.** Given that the dividend is five times more than the divisor, this means that the quotient is equal to 5. Given also that the divisor is twice more than the quotient, this means that the divisor is equal to 10. As the dividend is five times more than the divisor, so the dividend is equal to 50. Then, we get that the sum of the dividend, divisor, and quotient is $50 + 10 + 5 = 65$. The correct answer is (D).

14. At most, how many common squares can a 4×8 rectangle and a 3×9 rectangle have?

(A) 12   (B) 16   (C) 24   (D) 26   (E) 27

> Answer. (C)
> **Solution.** The common part of these two rectangles is a rectangle, so that its length is not more than 8 and its width is not more than 3. So, they can have at most $3 \cdot 8 = 24$ common squares (see the figure). The correct answer is (C).
>
>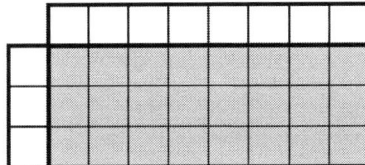

15. Which of these shapes is not possible to form using eight cubes?

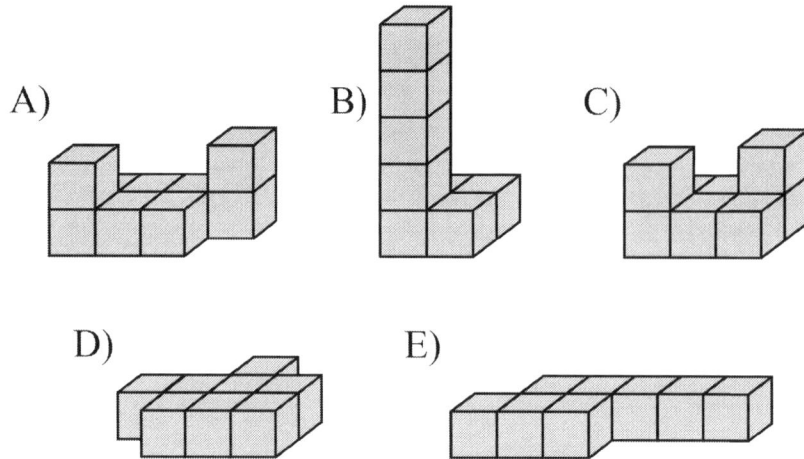

A) B) C) D) E)

Answer. (A)

Solution. Shapes of answers (B), (C), (D), (E) consist of eight cubes, while the shape of answer (A) consists of nine cubes. So, the shape of answer (A) is not possible to form using eight cubes. The correct answer is (A).

16. Eight numbers are written in one line. The first number is 2, the second number is 3, then starting from the third number each number is equal to the greatest prime that is less than or equal to the sum of the previous two numbers. For example, the third number must be equal to 5, because the sum of the first two numbers is $2 + 3 = 5$ and the third number is equal to the greatest prime that is less than or equal to 5 (that prime is 5). What is the eighth number?

(A) 37   (B) 31   (C) 29   (D) 17   (E) 13

Answer. (A)

Solution. We have that the first three numbers are:

$$2, 3, 5, \ldots$$

So, the forth number is the greatest number that is less than or equal to the sum of the previous two numbers. As $3 + 5 = 8$, then the greatest prime less than or equal to 8 is 7. So, the first four numbers are:

$$2, 3, 5, 7, \ldots$$

Continuing in a similar way, we get that these eight numbers are:

$$2, 3, 5, 7, 11, 17, 23, 37.$$

So, the eighth number is 37. The correct answer is (A).

17. A cubical aquarium of edge length 6 centimeters is full of water. All the water is poured out into another cubical aquarium of edge length 8 centimeters. What is the height (in centimeters) of the water?

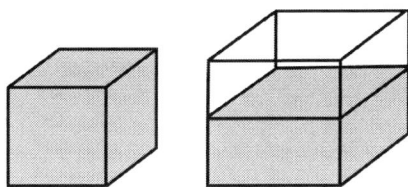

(A) 3   (B) 3.3   (C) 3.375   (D) 3.5   (E) 3.75

Answer. (C)

Solution. Let the height (in centimeters) of the water be x. So, when the water was in a cubical aquarium of edge length 6 centimeters its volume was $6 \cdot 6 \cdot 6$ cm^3 and when the water was in a cubical aquarium of edge length 8 centimeters its volume was $8 \cdot 8 \cdot x$. This means that

$$6 \cdot 6 \cdot 6 = 8 \cdot 8 \cdot x.$$

We get

$$x = \frac{6 \cdot 6 \cdot 6}{8 \cdot 8} = \frac{27}{8} = 3.375.$$

So, the height of the water is 3.375 centimeters. The correct answer is (C).

---

18. A pair (a, b) of natural numbers a and b is called "the simplest" if both $a - b$ and $a + b$ are prime numbers. If the pair (a, b) is "the simplest", then which of these numbers cannot be equal to $a - b$?

(A) 11   (B) 7   (C) 5   (D) 3   (E) 2

Answer. (E)

Solution. For example, the pairs $(12, 1)$, $(9, 2)$, $(6, 1)$, $(5, 2)$ are "the simplest", because both $12 - 1 = 11$ and $12 + 1 = 13$ are prime, both $9 - 2 = 7$ and $9 + 2 = 11$ are prime, both $6 - 1 = 5$ and $6 + 1 = 7$ are prime, and both $5 - 2 = 3$ and $5 + 2 = 7$ are prime. So, $a - b$ can be 11, 7, 5, 3. Note that $a - b$ cannot be 2, because in that case $a = b + 2$ and $a + b = b + 2 + b = 2b + 2$. As b is a natural number and $a + b = 2b + 2$, we get $a + b$ is an even number greater than 2. Then, $a + b$ is not a prime number. So, 2 cannot be equal to $a - b$. The correct answer is (E).

19. In every square of a 4×4 square is written one of the numbers 1, 2, 3, 4, so that in every row, in every column, and in every shaded part are written all four numbers 1, 2, 3, 4 (see the figure). What is $a + b$?

(A) 3   (B) 4   (C) 5   (D) 6   (E) 7

Answer. (C)
Solution. At first, let us fill in all the numbers that we can be sure about their positions (see the figure).

As in every column all four numbers 1, 2, 3, 4 must be written. Then, we get that $a + b = 2 + 3 = 5$. Here is an example of such a 4×4 square.

The correct answer is (C).

20. Four numbers are written on four pieces of paper (see the figure).

$$\boxed{15} \quad \boxed{2024} \quad \boxed{2022} \quad \boxed{1111}$$

Putting together some of these paper pieces we can get new numbers. For example

$$\boxed{2024} \quad \boxed{2022} \quad \boxed{1111} \quad \boxed{15}$$

How many numbers divisible by 15 can we get? You can use 1, 2, 3, or 4 pieces of paper.

(A) 4   (B) 6   (C) 8   (D) 10   (E) 11

Answer. (D)

Solution. A number is divisible by 15 if it is divisible by 3 and 5. A number is divisible by 5 if its last digit is either 5 or 0. So, 15 must be in the end. A number is divisible by 3 if and only if the sum of its digits is divisible by 3. So, we need to use the following pieces of paper:
$\boxed{2024}\,\boxed{2022}\,\boxed{1111}\,\boxed{15}$ or $\boxed{2024}\,\boxed{1111}\,\boxed{15}$ or $\boxed{2022}\,\boxed{15}$ or $\boxed{15}$. Using these pieces of papers we can get 1, 1, 2, 6 numbers, so that each number ends by 15 and is divisible by 15. So, altogether we can get $1 + 1 + 2 + 6 = 10$ numbers that are divisible by 15. The correct answer is (D).

Part C: Each correct answer is worth 5 points

21. A scale is broken and gives a reading that is 10 percent more than the actual weight of the item. It shows that a bag of pears weights 5.5 pounds. What is the actual weight (in pounds) of this bag of pears?

(A) 4.95   (B) 5   (C) 5.1   (D) 5.15   (E) 6

Answer. (B)

Solution. A scale is broken and gives a reading that is 10 percent more than the actual weight of the item. This means that the scale actually shows 100 percent + 10 percent of the actual weight. So, 110 percent of the actual weight is 5.5 pounds. If we divide the whole to 11 parts (each part is 10 percent), then each part is $\frac{5.5}{11} = 0.5$ pounds (see the figure). So, the actual weight of this bag of pears is 100 percent (that is 10 parts) and as each part is 0.5 pounds, then $10 \cdot 0.5 = 5$ pounds. The correct answer is (B).

22. A 2×2 paper square needs to be cut into four shapes (see the figure), so that each of these four shapes includes the number 1 and the number 2. In how many different ways is it possible to do this?

(A) 0   (B) 1   (C) 2   (D) 3   (E) 4

Answer. (C)

Solution. Each of these four shapes has one of the following forms (see the figure).

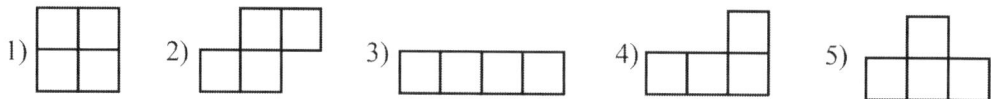

The correct answer is (C).

23. Five-digit number $\overline{aabbb}$ is divisible by 32. What is the greatest possible value of $a + b$?

(A) 10   (B) 15   (C) 16   (D) 17   (E) 18

Answer. (B)

Solution. Five-digit number $\overline{aabbb}$ we can represent in the following way:

$$\overline{aabbb} = 1000 \cdot \overline{aa} + \overline{bbb}.$$

Given that $\overline{aabbb}$ is divisible by 32, so it is divisible by 8. Note that 1000 is also divisible by 8. Then, from the last equation we get that $\overline{bbb}$ is divisible by 8. This means that either $b = 0$ or $b = 8$.

If $b = 8$, then to find the greatest possible value of $a + b$ we need to find the greatest possible value of a. Note that five-digit numbers 99888 and 88888 are not divisible by 8, while five-digit number 77888 is divisible by 32. So, the greatest possible value of a is 7 and the greatest possible value of $a + b$ is 15.

If $b = 0$, then as the digit a can be at most 9, we get $a + b \leq 9$.

So, the greatest possible value of $a + b = 15$. The correct answer is (B).

---

24. A 4×4 paper square and a 3×3 paper square each consist of 1×1 squares, the sides of which they can be cut along to form parts. What is the smallest possible number of parts needed to construct a 5×5 square?

(A) 3   (B) 4   (C) 5   (D) 6   (E) 7

Answer. (B)

Solution. Note that no two corner squares of four corner squares of a 5×5 can belong at the same time to any of these parts. So, we get that there must be at least four parts. Here is an example of a 5×5 square that was created by four such parts.

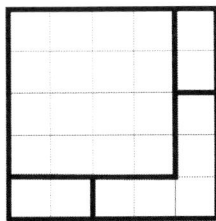

So, the smallest number of such parts is four. The correct answer is (B).

25. A four-digit number is called "interesting" if exactly three of its digits are the same. How many four-digit "interesting" numbers are there so that for each of them the next number is also an "interesting" number?

(A) 8   (B) 17   (C) 72   (D) 73   (E) 81

Answer. (C)

Solution. Four-digit "interesting" numbers that start with the digit 1 are:

$$1112,\ 1113,\ 1114,\ 1115,\ 1116,\ 1117,\ 1118,\ \text{and } 1999.$$

Four-digit "interesting" numbers that start with the digit 2 are:

$$2220,\ 2223,\ 2224,\ 2225,\ 2226,\ 2227,\ 2228,\ \text{and } 2999.$$

Four-digit "interesting" numbers that start with the digit 3 are:

$$3330,\ 3331,\ 3334,\ 3335,\ 3336,\ 3337,\ 3338,\ \text{and } 3999.$$

Four-digit "interesting" numbers that start with the digit 4 are:

$$4440,\ 4441,\ 4442,\ 4445,\ 4446,\ 4447,\ 4448,\ \text{and } 4999.$$

Four-digit "interesting" numbers that start with the digit 5 are:

$$5550,\ 5551,\ 5552,\ 5553,\ 5556,\ 5557,\ 5558,\ \text{and } 5999.$$

Four-digit "interesting" numbers that start with the digit 6 are:

$$6660,\ 6661,\ 6662,\ 6663,\ 6664,\ 6667,\ 6668,\ \text{and } 6999.$$

Four-digit "interesting" numbers that start with the digit 7 are:

$$7770,\ 7771,\ 7772,\ 7773,\ 7774,\ 7775,\ 7778,\ \text{and } 7999.$$

Four-digit "interesting" numbers that start with the digit 8 are:

$$8880,\ 8881,\ 8882,\ 8883,\ 8884,\ 8885,\ 8886,\ \text{and } 8999.$$

Four-digit "interesting" numbers that start with the digit 9 are:

$$9990,\ 9991,\ 9992,\ 9993,\ 9994,\ 9995,\ 9996,\ 9997.$$

So, there are 72 four-digit "interesting" numbers are there so that for each of them the next number is also an "interesting" number. The correct answer is (C).

26. A *cryptarithm* is a mathematical puzzle where the digits have been replaced by letters. How many solutions does the following cryptarithm have?

$$\overline{AR} \times \overline{TS} = \overline{AKH}.$$

(A) 0   (B) 1   (C) 2   (D) 3   (E) 4

Answer. (A)
Solution. $T > 1$ is not possible, because if $T > 1$, then the first digit of the product $\overline{AR} \times \overline{TS}$ will be greater than A and it cannot be equal to $\overline{AKH}$. So $T = 1$. Note that $A \geq 2$, $R \geq 2$, $S \geq 2$ and $R + A \cdot S \leq 9$. On the other hand $R + A \cdot S \geq 4 + 2 \cdot 3 = 10$. This is not possible. So, this cryptarithm has no solutions. The correct answer is (A).

---

27. Using exactly five of these six paper shapes a paper square was created (none of these shapes can have an overlapping region). Which shape was not used?

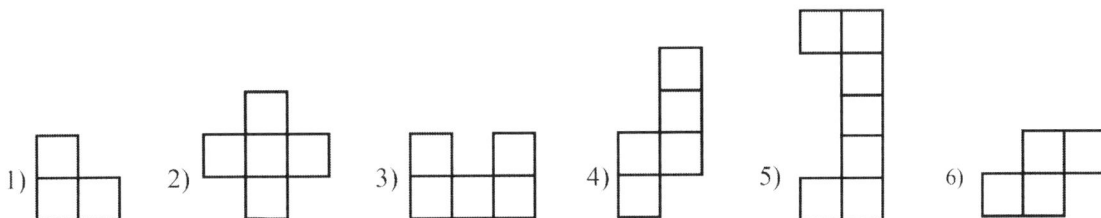

(A) 1   (B) 2   (C) 3   (D) 5   (E) 6

Answer. (E)
Solution. If the area of each little square is for example 1, then the areas of these shapes are 3, 5, 5, 5, 7, 4. So, if we choose exactly five of these six shapes to put them together and create a new shape, then its area can be at least 22 (as $22 = 3 + 5 + 5 + 5 + 4$) and at most 26 (as $26 = 5 + 5 + 5 + 7 + 4$). As it needs to be a square, then its area must be 25. We can get 25 if we choose the shapes with areas 3, 5, 5, 5, 7. So, the shape number 6) was not used. Here is an example of a square that can be created using the first five shapes (see the figure).

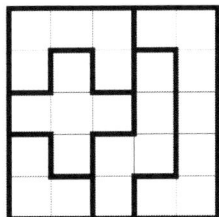

The correct answer is (E).

---

28. Two groups of ducks are swimming in the river. The number of ducks in the first group is a two-digit number $\overline{ab}$, the number of ducks in the second group is a two-digit number $\overline{ba}$. Some ducks moved from the first group to the second group, after this the number of ducks in the first group became a two-digit number $\overline{cd}$ and the number of ducks in the second group became a two-digit number $\overline{dc}$, where c and d may be the same, but $b \neq d$. At least how many ducks could move from the first group to the second group?

(A) 5   (B) 6   (C) 8   (D) 9   (E) 10

Answer. (D)
Solution. The overall number of ducks did not change, so we have $\overline{ab} + \overline{ba} = \overline{cd} + \overline{dc}$. This means that $10 \cdot a + b + 10 \cdot b + a = 10 \cdot c + d + 10 \cdot d + c$. We get $a + b = c + d$. The number of ducks that moved from the first group to the second group is $\overline{ab} - \overline{cd} = 10 \cdot (c - d) + b - d = 9 \cdot (d - b)$. This means that the number of ducks that moved from the first group to the second group is divisible by 9 (so it is not less than 9, as $b - d \neq 0$). We provide an example for 9. For example, $\overline{ab} = 31$, $\overline{ba} = 13$, $\overline{cd} = 22$, $\overline{dc} = 22$, then the number of ducks that moved from the first group to the second group is $31 - 22 = 9$. The correct answer is (D).

29. At least how many times a 6×6 square must be folded to get a 1×1 paper square?

(A) 5   (B) 6   (C) 7   (D) 8   (E) 9

Answer. (B)
Solution. Let us color one of the squares of this 6×6 paper square in a different color (for example in black). Assume that when we fold this paper square and the black 1×1 square touches any other 1×1 square, then that 1×1 square also becomes black. So, when the paper is folded once the number of black 1×1 squares is not more than 2, when the paper is folded twice the number of black 1×1 squares is not more than 4, when the paper is folded three times the number of black 1×1 squares is not more than 8, when the paper is folded four times the number of black 1×1 squares is not more than 16, and when the paper is folded five times the number of black 1×1 squares is not more than 32. Then, in order this 6×6 square to be folded to get a 1×1 square, all its 1×1 squares (36 such squares) must be folded to get one 1×1 square. So, in order a 6×6 square to be folded to get a 1×1 square it needs to be folded at least six times. Here is an example showing that if a 6×6 square is folded six times then one can get a 1×1 square (see the figure).

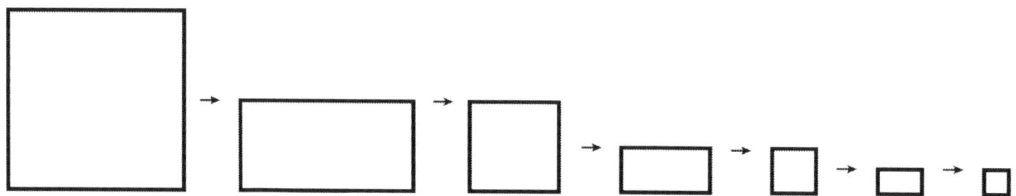

The correct answer is (B).

30. The numbers 1, 2, 3, 4, 5, 6 are written on the faces of a cube (one number per face). For any two faces that share a common edge the positive difference of the numbers written on these two faces is written on that edge. What is the smallest possible sum of all numbers written on all 12 edges?

(A) 23   (B) 24   (C) 25   (D) 26   (E) 27

Answer. (D)

Solution. Let this sum be S. If we replace 1 by 2, then the sum S will be replaced by $S - 4$. If we replace 6 by 5, then $S - 4$ will be replaced by $S - 8$. This means that the new numbers written on the faces of the cube are 2, 2, 3, 4, 5, 5, and the sum of all numbers written on all 12 edges is $S - 8$. Note that from these 12 numbers at most two of them can be equal to 0, at most five of them can be equal to 1, and at most four of them can be equal to 2. Moreover, if five of them are equal to 1, then four of them cannot be equal to 2, otherwise we get that the face with number 3 has a common edge with five different faces. This is impossible. Then, we get

$$S - 8 \geq 0 + 0 + 1 + 1 + 1 + 1 + 1 + 2 + 2 + 2 + 3 + 3.$$

So, we get $S \geq 25$, and note that S must be an even number, so $S \geq 26$. We provide an example for $S = 26$.

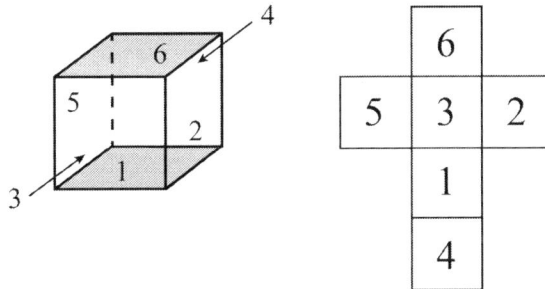

The correct answer is (D).

Solutions of Test 3

1. Beach chairs and umbrellas are ordered in one line like this.

There are 50 umbrellas, at most how many chairs can there be?

(A) 50   (B) 52   (C) 100   (D) 101   (E) 102

> Answer. (C)
> **Solution.** Each umbrella is in between of two chairs, so there can be at most $50 \cdot 2 = 100$ chairs. The correct answer is (C).

2. Sam has two watches, one is 5 minutes behind and the other one is 10 minutes ahead. One of the watches shows that it is 10:27 am, the other one shows that it is 10:12 am. What time is it?

(A) 10:02 am   (B) 10:22 am   (C) 10:17 am   (D) 10:37 am   (E) cannot be determined

> Answer. (C)
> **Solution.** The watch that is 10 minutes ahead must show 10:27 am, so it is 10:17 am. The correct answer is (C).

3. A paper rectangle was folded across a line parallel to one of its sides. Then, the folded paper was cut by a straight line. Which of these parts is not possible to get?

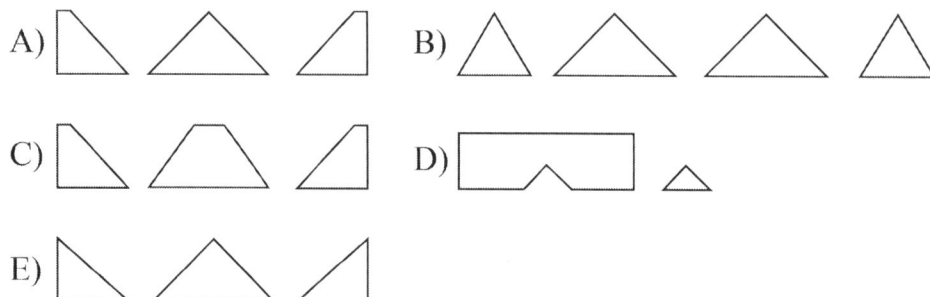

84

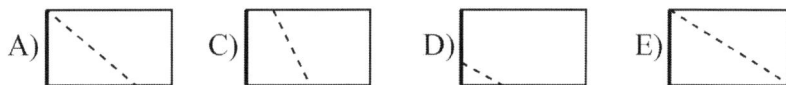
4. What is the sum of the digits of the smallest two-digit number $\overline{ab}$ that satisfies the following inequality $\overline{ab} < (a+1) \cdot (b+1)$?

(A) 7   (B) 8   (C) 9   (D) 10   (E) 18

Answer. (D)

Solution. The smallest two-digit number is 10, so we start with 10. We have

$$10 > 2 \cdot 1.$$

So, for 10 this inequality $\overline{ab} < (a+1) \cdot (b+1)$ is not correct. We continue

$$11 > 2 \cdot 2, \quad 12 > 2 \cdot 3, \quad 13 > 2 \cdot 4, \quad 14 > 2 \cdot 5, \quad 16 > 2 \cdot 7, \quad 18 = 2 \cdot 9, \quad 19 < 2 \cdot 10.$$

This means that the smallest two-digit number $\overline{ab}$ that satisfies the following inequality $\overline{ab} < (a+1) \cdot (b+1)$ is 19. Then, the sum of its digits is $1 + 9 = 10$. The correct answer is (D).

5. At 11:30 am two groups of ducks were swimming in the river. Starting from noon till 12:17 pm, each 3 minutes 2 ducks moved from the first group to the second group, and every 5 minutes 3 ducks moved from the second group to the first group. Given that 12:16 pm there were 9 ducks in the first group. How many ducks were in the first group at 12:01 pm?

(A) 12   (B) 11   (C) 19   (D) 17   (E) 8

Answer. (C)

Solution. From 12:01 pm to 12:16 pm $5 \cdot 2 = 10$ ducks left the first group, and at 12:16 pm there were 9 ducks in the first group. This means that at 12:01 pm there were $9 + 10 = 19$ ducks in the first group. The correct answer is (C).

6. How many digits are there, so that there are at least two different ways to move one stick to get another digit?

(A) 0   (B) 1   (C) 2   (D) 3   (E) 4

Answer. (C)

Solution. The digits 0, 1, 2,..., 9 are written using 6, 2, 5, 5, 4, 5, 6, 3, 7, 6 sticks. So, in order to have at least two different ways to move one stick to get another digit it must be written either by 7 or by 6 sticks. The digits 0 and 6 are written by 6 sticks but they do not work. The digits 8 and 9 (are written by 7 and 6 sticks) and there are two different ways to move one stick to get another digit, here is an example.

So, there are two digits (8 and 9) satisfying the conditions. The correct answer is (C).

7. Each circle must be filled in by a different odd digit so that all inequalities are correct. What digit must be in the rightmost circle?

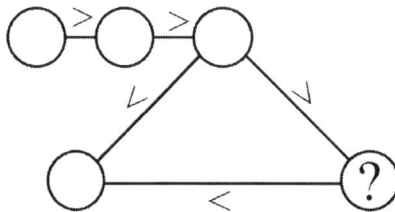

(A) 9   (B) 7   (C) 5   (D) 3   (E) 1

Answer. (D)

Solution. The greatest and the smallest odd digits must be written like this:

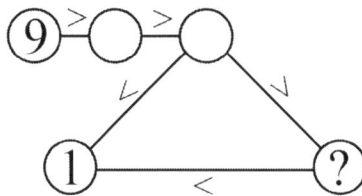

All other odd digits can be filled in one single way, like this:

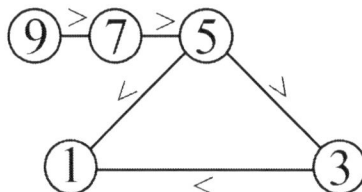

So, in the rightmost circle must be the digit 3. The correct answer is (D).

8. Five coins are placed in a row on a table. In how many different ways is possible to turn two coins? It is not allowed to turn any neighbor coins.

$$\bigcirc 10¢ \; \bigcirc 10¢ \; \bigcirc 10¢ \; \bigcirc 10¢ \; \bigcirc 10¢$$

(A) 10   (B) 6   (C) 5   (D) 4   (E) 3

> **Answer. (B)**
> **Solution.** Turned coins are shaded. If the first coin is turned, then there are 3 cases. If the second coin is turned, there are 2 cases. If the third coin is turned, there is 1 case. If the fourth or fifth coin is turned, we do not get any new cases. There are $3+2+1 = 6$ ways.

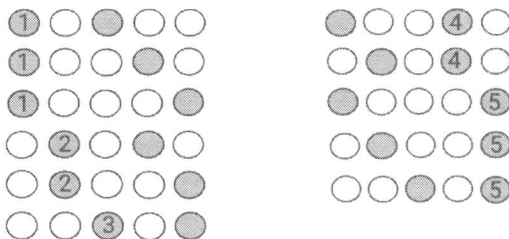

9. Six paper cards are placed in a row on a table. A number is written on each card, and three cards are face-up (as shown).

$$\boxed{4}\;\boxed{}\;\boxed{}\;\boxed{}\;\boxed{8}\;\boxed{9}$$

Given that the numbers on any three consecutive cards sum to the same number, what is the sum of the numbers on the three face-down cards?

(A) 21   (B) 17   (C) 16   (D) 12   (E) 10

> **Answer. (A)**
> **Solution.** Given that the numbers on any three consecutive cards sum to the same number, so the same number must be written on any two cards, so that in between of them there are exactly two cards. Then, the numbers on the face down cards are 8, 9, 4. So, the sum of the numbers on three face down paper cards is $8 + 9 + 4 = 21$. The correct answer is (A).

10. How many different rectangular prisms is possible to construct using eight unit cubes?

(A) 1   (B) 2   (C) 3   (D) 4   (E) 6

> **Answer. (C)**
> **Solution.** Unit cube means: width = length = height = 1. So, the width, length, and height of any rectangular prism we get from these eight cubes must be natural numbers and $width \times length \times height = 8$. Then, all possible rectangular prisms are $1 \times 1 \times 8$, $1 \times 2 \times 4$, and $2 \times 2 \times 2$. So, there are 3 different such rectangular prisms. The correct answer is (C).

11. The numbers 1, 1, 2, 2,..., 9, 9 are written on the blackboard. At most how many pairs of numbers is possible to form from these 18 numbers, so that the positive difference of the numbers in each pair is 1 or 2?

(A) 5   (B) 6   (C) 7   (D) 8   (E) 9

> Answer. (E)
> **Solution.** We can form at most 9 such pairs, for example
>
> $$(1,2),(2,3),(3,1),(4,5),(4,5),(6,7),(6,7),(8,9),(8,9).$$
>
> The correct answer is (E).

12. How many times 2 appears from 101 to 301?

(A) 20   (B) 40   (C) 100   (D) 120   (E) 140

> Answer. (E)
> **Solution.** As a units digit 2 appears $29 - 9 = 20$ times. As a tens digit 2 appears $2 \cdot 10 = 20$ times. As hundreds digit 2 appears 100 times. So, altogether 2 appears 140 times. The correct answer is (E).

13. Which of the following statements is correct?
(A) point M is outside the square and the triangle
(B) point M is outside the triangle and the circle
(C) point M is outside the circle or the square
(D) point M is inside the circle and the square
(E) point M is outside the circle and the square

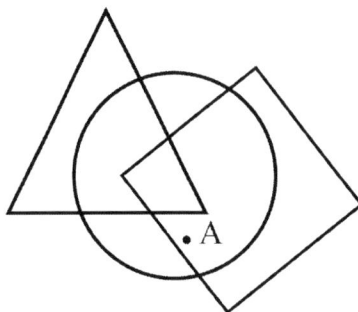

> Answer. (D)
> **Solution.** Point M is outside the triangle, inside the circle and inside the square. The correct answer is (D).

14. At least how many points need to be erased, so that no three leftover points are on the same line (see the figure)?

(A) 1   (B) 2   (C) 3   (D) 4   (E) 5

Answer. (D)
Solution. Note that at least one point of each of these four rays needs to be erased (see the figure).

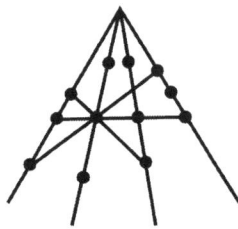

Note also that erasing four points is enough (see the figure).

So, at least four points need to be erased, so that no three leftover points are on the same line. The correct answer is (D).

15. Mia needed to solve her math homework problems in six days. The first day she solved $\frac{1}{8}$ part of her homework, the next four days she solved three problems per day, the sixth day she solved the last two problems. How many problems did she solve the first day?

(A) 1   (B) 2   (C) 3   (D) 4   (E) 5

Answer. (B)
Solution. Given that the first day Mia solved $\frac{1}{8}$ part of her homework problems, so the other five days she was left with $\frac{7}{8}$ part of her homework problems. Given also that in these five days she solved $4 \cdot 3 + 2 = 14$ problems. So, $\frac{7}{8}$ part of her homework problems is equal to 14. This means that $\frac{1}{8}$ part of her homework problems is 7 times less than 14. Thus, $\frac{1}{8}$ part of her homework problems is 2. So, the first day she solved 2 problems. The correct answer is (B).

16. An elevator in a 9-story building has it's stops labeled on a digital screen as follows:

$$\mathsf{M,1,2,3,4,5,6,7,8}$$

Mia was looking at the elevator's mirror and looked at the reflection of the screen and saw the number of the floor where she lives. When she left the elevator, she realized she had gotten off too early. On what floor does Mia live?

(A) 1   (B) 2   (C) 3   (D) 5   (E) 8

> **Answer. (D)**
> **Solution.** As she was the number in the mirror, then it means that this number is the symmetric number of the number that the digital screen showed. Given also that Mia went out earlier than needed, so she lives on 5th floor and she went out on the 2nd floor. The correct answer is (D).

17. On the first day a car travelled $\frac{1}{3}$ part of the distance on a straight road, on the second day it travelled $\frac{1}{3}$ part of the remaining road, on the third day it travelled $\frac{1}{3}$ part of the remaining road, and on the fourth day it travelled all the remaining part of the road. What part of the entire road was travelled on the fourth day?

(A) $\frac{1}{3}$ (B) $\frac{2}{9}$ (C) $\frac{4}{27}$ (D) $\frac{8}{27}$ (E) $\frac{1}{4}$

> **Answer. (D)**
> **Solution.** On the first day the car travelled $\frac{1}{3}$ part of the entire road. As on the second day the car travelled $\frac{1}{3}$ part of the remaining road, then on the second day it travelled $\left(1-\frac{1}{3}\right)\cdot\frac{1}{3}=\frac{2}{9}$. As on the third day the car travelled $\frac{1}{3}$ part of the remaining road, then on the third day it travelled $\left(1-\frac{1}{3}-\frac{2}{9}\right)\cdot\frac{1}{3}=\frac{4}{27}$ part of the entire road. As on the fourth day the car travelled all the remaining part of the road, then on the fourth day it travelled $1-\frac{1}{3}-\frac{2}{9}-\frac{4}{27}=\frac{8}{27}$ part of the entire road. The correct answer is (D).

18. A pentagon is divided by one of its diagonals into a triangle and a quadrilateral. The perimeters of a triangle, quadrilateral, and pentagon are 13, 14, 15, respectively. What is the length of that diagonal?

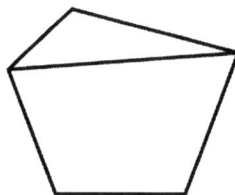

(A) 5   (B) 6   (C) 6.5   (D) 7   (E) 7.5

Answer. (B)
Solution. If the length of this diagonal is x, then note that the sum of the triangle's and quadrilateral's perimeters is equal to the perimeter of the pentagon plus $2 \cdot x$. So, we get

$$2x = 13 + 14 - 15.$$

Thus, it follows that

$$x = \frac{13 + 14 - 15}{2} = 6.$$

The length of the diagonal is 6. The correct answer is (B).

---

19. Sam and Luna had the same math homework problems to solve. Each day Sam solved three problems, and Luna solved four problems. When Luna was done with her entire math homework Sam needed to work four more days to be done. From how many problems did the math homework consist of?

(A) 20   (B) 24   (C) 36   (D) 48   (E) 68

Answer. (D)
Solution. When Luna was done with her entire math homework Sam needed to solve $4 \cdot 3 = 12$ more problems. Each day Luna solved one more problem than Sam, this means that Luna has spent 12 days to be done with her entire math homework. As each day Luna solved four problems, then the entire math homework consisted of $12 \cdot 4 = 48$ problems. The correct answer is (D).

---

20. Seven coins are placed in a row on a table. In how many different ways is possible to turn three coins? It is not allowed to turn any neighbor coins.

(A) 7   (B) 8   (C) 9   (D) 10   (E) 11

Answer. (B)
Solution. Turned coins are shaded. If the first coin is turned, then there are 6 cases (see the left part). If the second coin is turned, there are 3 cases (right part). If the third coin is turned, there is 1 case (right part). If the fourth coin is turned, we get the same cases as before (one can check this by shading the last two rows of blank coins). The same for the fifth and sixth coins. So, the number of all ways is $6 + 3 + 1 = 10$. The correct answer is (D).

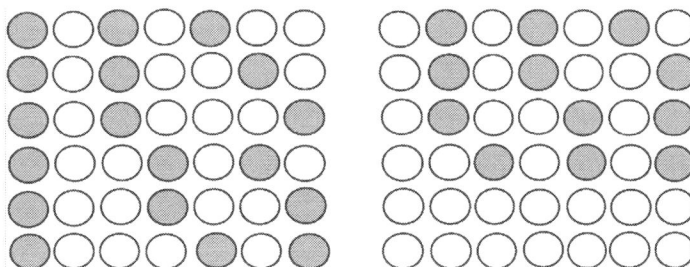

Part C: Each correct answer is worth 5 points

21. How many four-digit numbers are there that can be written using each of the following digits 1, 2, 3, 4, so that any two neighbor digits are of different parity?

(A) 16   (B) 20   (C) 24   (D) 32   (E) 64

Answer. (D)
Solution. There are four different ways to choose the first digit of this four-digit number. As the parity of the next digit must be different, then there are two different ways to choose the next digit. In a similar way, there are two different ways to choose the third and there are two different ways to choose the fourth digit. So, the number of all such four-digit numbers is $4 \cdot 2 \cdot 2 \cdot 2 = 32$. The correct answer is (D).

22. There are apples and pears in a fruit bowl. The number of apples is 40 percent of all apples and pears in the bowl. Mia ate one of the apples and the number of apples became equal to 50 percent of all pears in the bowl. Then, Bob ate one of the pears. What percent of the pears are the apples?

(A) 60   (B) 55   (C) 50   (D) 40   (E) 33

Answer. (A)
Solution. As the number of apples was 40 percent of all apples and pears in the bowl. So, if there were $4x$ apples, then the number of pears was $6x$. Mia ate one of the apples and the number of apples became equal to 50 percent of all pears in the bowl. This means that

$$4x - 1 = 6x \cdot \frac{1}{2}.$$

We get $x = 1$. So, there were 4 apples and 6 pears in the bowl. Mia ate one apple, so 3 apples were left. Bob ate one pear, so 5 pears were left. Finding what percent of the pears are the apples, means finding this

$$\frac{3}{5} \cdot 100\% = 60\%.$$

The correct answer is (A).

23. There are 10 chairs next to a round table. At least how many people need to sit on these chairs (each chair can be used only by one person), so that for each chair at least one of its neighbor chairs is occupied by someone?

(A) 3   (B) 4   (C) 5   (D) 6   (E) 7

Answer. (D)

Solution. Note that if a chair is used by someone, then at least one of its neighbor chairs is also used by someone. So, the number of people must be greater than 5. Because if there are not more than five people, then there are three consecutive unused chairs. This is not possible. We provide an example, where 6 people can sit on the chairs, so that for each chair at least one of its neighbor chairs is occupied by someone.

The correct answer is (D).

24. The common part of a rectangle of perimeter 6 and a rectangle of perimeter 8 is a rectangle of a perimeter 3 (see the figure). What is the perimeter of rectangle $PQRS$?

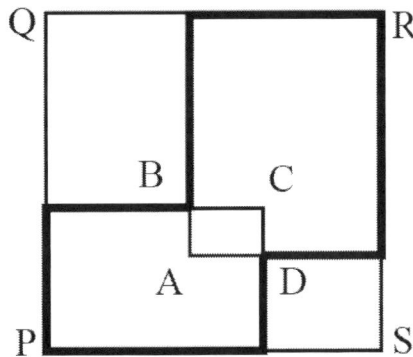

(A) 7   (B) 9   (C) 10   (D) 11   (E) 12

Answer. (D)

Solution. From the figure we can easily see that the sum of the perimeters of rectangles of perimeters 6 and 8 is equal to the sum of the perimeters of rectangles $ABCD$ and $PQRS$. So, the perimeter of rectangle $PQRS$ is equal to $6 + 8 - 3 = 11$. The correct answer is (D).

25. A floor is tiled with 20×40 tiles. An ant starts to move from point A to point B, it can move only along the sides of the tiles and it chooses the shortest route. In how many different ways can the ant go from A to B?

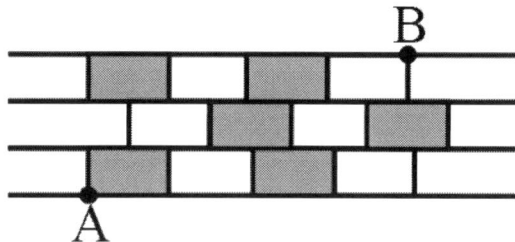

(A) 8   (B) 10   (C) 13   (D) 19   (E) 20

Answer. (E)

Solution. Let us consider all vertices in between of A and B. At each vertex let as write the number of shortest routes going from A to that vertex (see the figure).

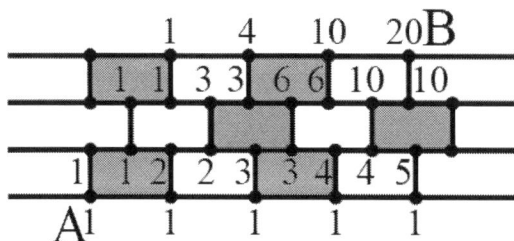

The figure shows that the ant can go from A to B in 20 such ways. The correct answer is (E).

26. Luna needs to fry 5 hamburger patties. It takes 2 minutes to fry each side of each patty. At most 4 patties fit to her kitchen pan. At least in how many minutes can Luna fry all 5 patties?

(A) 4   (B) 5   (C) 6   (D) 7   (E) 8

Answer. (B)

Solution. Assume that each patty is divided to two patties (each one twice thinner than the initial one). So, we have 10 such patties. In 2 minutes Luna can fry 4 such patties. Luna needs $2 \cdot \frac{10}{4} = 5$ minutes to fry all 10 patties. We provide an example how can she do it in 5 minutes. So, Luna can fry all 5 patties at least in 5 minutes. The correct answer is (B).

27. How many five-digit numbers $\overline{abcde}$ are there, so that $b = a+1$, $c = b+1$, $e = d+1$?

(A) 63   (B) 56   (C) 50   (D) 16   (E) 10

Answer. (A)

Solution. The are 7 such three-digit numbers $\overline{abc}$, namely 123, 234, 345, 456, 567, 678, 789. The digits d and e we can choose in 9 different ways, namely (0, 1), (1, 2), (2, 3), (3, 4), (4, 5), (5, 6), (6, 7), (7, 8), (8, 9). So, there are $7 \cdot 9 = 63$ such five-digit numbers $\overline{abcde}$. The correct answer is (A).

---

28. At most in how many squares of a 8×8 square is possible to put ★, so that there is at most one ★ in any such ⌐| shape (in any position: rotated or flipped)?

(A) 8   (B) 9   (C) 10   (D) 11   (E) 12

Answer. (C)

Solution. Note that if you put ★ in two different squares of a 2×3 rectangle, then they belong to the same ⌐| shape. Taking this into consideration and the figure below

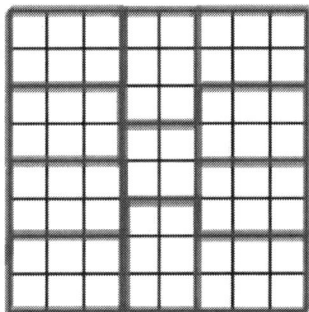

we get that there are at most ten ★ (as there are ten 2×3 rectangles).

Let us provide an example of ten such ★ (see the figure below).

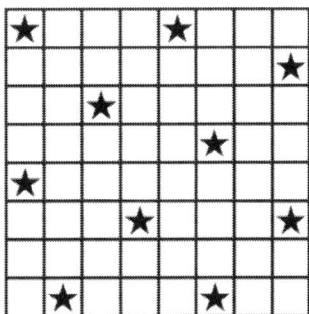

The correct answer is (C).

---

29. At least in how many colors can the squares of 8×8 square be painted, so that any 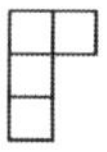 shape (in any position: rotated or flipped) does not include two or more squares of the same color?

(A) 7　　　　(B) 8　　　　(C) 9　　　　(D) 10　　　　(E) 11

Answer. (B)

Solution. At first, let us prove that we cannot use 7 or less colors. Note that all seven squares of the bold shape must be of different colors (see the figure). Then, the colors of all other squares can be found (uniquely).

This is not possible, as there is ⌐ shape that includes two squares of the same color (see the figure below, two circled 3).

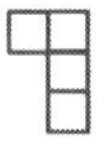

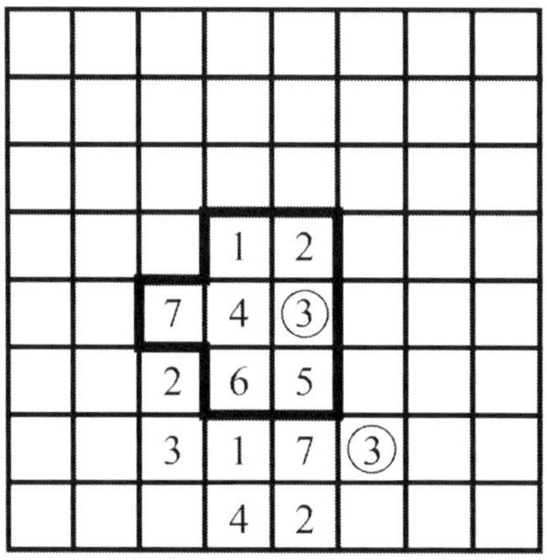

		1	2				
	7	4	③				
	2	6	5				
	3	1	7	③			
		4	2				

We provide an example for 8 different colors.

5	8	3	2	5	8	3	2
7	6	1	4	7	6	1	4
3	2	5	8	3	2	5	8
1	4	7	6	1	4	7	6
5	8	3	2	5	8	3	2
7	6	1	4	7	6	1	4
3	2	5	8	3	2	5	8
1	4	7	6	1	4	7	6

The correct answer is (B).

30. In how many different ways can we fill in the squares by the numbers 1, 2, 3, 4, 5, 6, 7, so that the sum of all four numbers of each circle is the same number? Only one number can be written in a square and all numbers must be used.

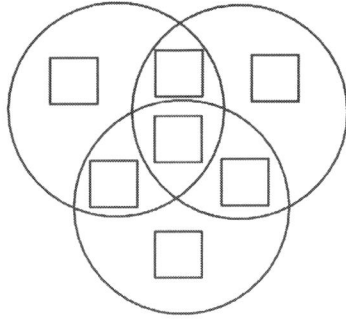

(A) 60   (B) 100   (C) 110   (D) 108   (E) 120

Answer. (D)

Solution. Let us write in the squares the numbers a, b, c, d, e, f, g, so that they are all different and each of them is a number from 1 to 7.

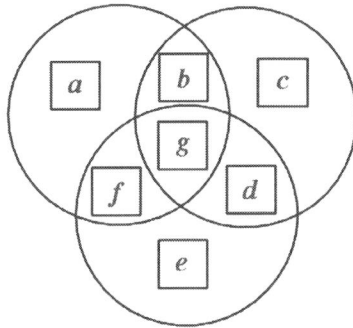

Given that the sum of all four numbers of each circle is the same number, so

$$a + b + g + f = c + b + d + g = e + d + f + g.$$

From here, we get

$$a - d = c - f = e - b.$$

Let $a - d = c - f = e - b = k$, note that here k can be $-4, -3, -2, -1, 1, 2, 3, 4$.
If $k = 4$ or $k = -4$, then we can fill in the squares in $3 \cdot 2 \cdot 1 = 6$ different ways.
If $k = 3$ or $k = -3$, then we can fill in the squares in $2 \cdot 3 \cdot 2 \cdot 1$ different ways.
If $k = 2$ or $k = -2$, then we can fill in the squares in $2 \cdot 3 \cdot 2 \cdot 1$ different ways.
If $k = 1$ or $k = -1$, then we can fill in the squares in $4 \cdot 3 \cdot 2 \cdot 1$ different ways.
So, altogether there are

$$2 \cdot (3 \cdot 2 \cdot 1 + 2 \cdot 3 \cdot 2 \cdot 1 + 2 \cdot 3 \cdot 2 \cdot 1 + 4 \cdot 3 \cdot 2 \cdot 1) = 2 \cdot 9 \cdot 3 \cdot 2 \cdot 1 = 108$$

different ways to fill in the squares. The correct answer is (D).

Solutions of Test 4

Part A: Each correct answer is worth 3 points

1. A math teacher gave students five problems. Each student solved either three, four, or five problems. Two students solved five problems. The number of students who solved four problems is 3 times more than the number of students who solved five problems and is 2 times less than the number of students who solved three problems. How many students are there?

(A) 18   (B) 19   (C) 20   (D) 22   (E) 24

> **Answer.** (C)
> **Solution.** Given that two students solved five problems and three times as many students solved four problems as solved five problems, so $2 \cdot 3$ students solved four problems. Given also that half as many students solved four problems as solved three problems, so $2 \cdot 3 \cdot 2$ students solved four problems. So, there are two students who solve five problems, six students who solved four problems and 12 students who solved three problems. As $2 + 6 + 12 = 20$, then altogether there are 20 students. The correct answer is (C).

2. For example, a rectangle has two *lines of symmetry* (see the picture).

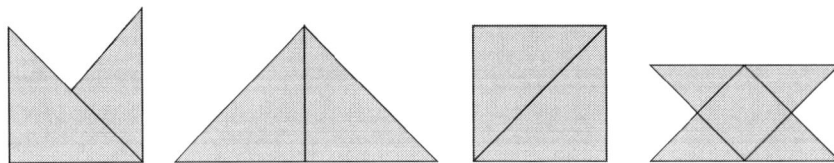

Match the number of lines of symmetry to each shape. Which answer is extra?

(A) 0   (B) 1   (C) 2   (D) 3   (E) 4

> **Answer.** (D)
> **Solution.** The first shape (from left to right) has 0 lines of symmetry, so it matches to answer (A).
> The second shape has has 1 line of symmetry, so it matches to answer (B).
> The third shape has has 4 lines of symmetry, so it matches to answer (E).
> The fourth shape has has 2 lines of symmetry, so it matches to answer (C).
> So, answer (D) is extra. The correct answer is (D).

3. In a five-day school week a student needed to solve in average six problems per day. The student solved seven math problems in the first day, seven math problems in the second day, four math problems in the third day, and five math problems in the fourth day. How many problems are left to be solved in the fifth day?

(A) 4   (B) 5   (C) 6   (D) 7   (E) 8

Answer. (D)
Solution. As the student needs to solve in average six problems per day (in a five-day school week), then altogether the student needs to solve $6 \cdot 5 = 30$ problems.
Given that in the first four days the student solved $7 + 7 + 4 + 5 = 23$ problems. So, there are $30 - 23 = 7$ problems left to be solve in the fifth day. The correct answer is (D).

4. The age of a girl in months is equal to the age of her grandmother in years. If the sum of their ages is 65 years, what is the (positive) age difference in years between them?

(A) 40   (B) 45   (C) 50   (D) 55   (E) 56

Answer. (D)
Solution. Let the girl be n years old. This means, she is $12 \cdot n$ months old. As the age of a girl in months is equal to the age of her grandmother in years, then the grandmother is $12 \cdot n$ years old.
Given also that the sum of their ages is 65 years, so

$$n + 12n = 65.$$

We get that

$$n = 5.$$

As n is the age of the girl and $12 \cdot n$ is the age of the grandmother, then the girl is 5 years old and the grandmother is 60 years old.
So, their (positive) age difference in years is:

$$60 - 5 = 55.$$

The correct answer is (D).

5. At most, how many of these five symbols $=, <, >, \leq, \geq$ can we use instead of ♣?

$$\frac{1}{2} + \frac{1}{3} + \frac{1}{5} + \frac{1}{6} \quad ♣ \quad 1 + \frac{1}{5}.$$

(A) 1   (B) 2   (C) 3   (D) 4   (E) 5

Answer. (C)

Solution. Both sides have $\frac{1}{5}$, so to compare both side we only need to compare the following

$$\frac{1}{2} + \frac{1}{3} + \frac{1}{6} \quad ♣ \quad 1.$$

Bringing these three fractions to a common denominator, we have

$$\frac{1}{2} + \frac{1}{3} + \frac{1}{6} = \frac{3}{6} + \frac{2}{6} + \frac{1}{6} = 1.$$

We get

$$\frac{1}{2} + \frac{1}{3} + \frac{1}{5} + \frac{1}{6} = 1 + \frac{1}{5}.$$

This means, if instead of ♣ we use $\leq$, then

$$\frac{1}{2} + \frac{1}{3} + \frac{1}{5} + \frac{1}{6} \leq 1 + \frac{1}{5},$$

is not actually a wrong statement (as the case of equality is true). In the same way, if instead of ♣ we use $\geq$, then

$$\frac{1}{2} + \frac{1}{3} + \frac{1}{5} + \frac{1}{6} \geq 1 + \frac{1}{5}.$$

is not actually a wrong statement (as the case of equality is true).

So, instead of ♣ we can use either $=, \leq$ or $\geq$. In other words, three of given five symbols can be used instead of ♣. The correct answer is (C).

6. Which of these rectangular prisms can be constructed using at most 17 unit cubes?

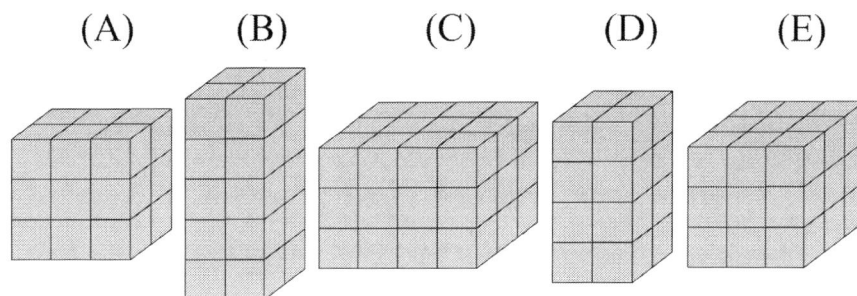

(A) (B) (C) (D) (E)

Answer. (D)

Solution. The dimensions of the rectangular prism of answer (A) are $3 \times 3 \times 2$, so we need $3 \cdot 3 \cdot 2 = 18$ unit cubes to construct it. This means, it cannot be constructed using at most 17 unit cubes (as $18 > 17$).

The dimensions of the rectangular prism of answer (B) are $2 \times 2 \times 5$, so we need $2 \cdot 2 \cdot 5 = 20$ unit cubes to construct it. This means, it cannot be constructed using at most 17 unit cubes (as $20 > 17$).

The dimensions of the rectangular prism of answer (C) are $2 \times 3 \times 4$, so we need $2 \cdot 3 \cdot 4 = 24$ unit cubes to construct it. This means, it cannot be constructed using at most 17 unit cubes (as $24 > 17$).

The dimensions of the rectangular prism of answer (D) are $2 \times 2 \times 4$, so we need $2 \cdot 2 \cdot 4 = 16$ unit cubes to construct it. This means, it can be constructed using at most 17 unit cubes (as $16 < 17$).

The dimensions of the rectangular prism of answer (D) are $3 \times 3 \times 3$, so we need $3 \cdot 3 \cdot 3 = 27$ unit cubes to construct it. This means, it cannot be constructed using at most 17 unit cubes (as $27 > 17$). The correct answer is (D).

7. The product of three different natural numbers is equal to 30. Which of the following answers cannot be equal to the sum of these three numbers?

(A) 10   (B) 12   (C) 14   (D) 16   (E) 18

Answer. (D)

Solution. Let us write the number 30 as a product of three different natural numbers in the following four ways.

$$30 = 1 \cdot 2 \cdot 15, \text{ note that } 1 + 2 + 15 = 18.$$

$$30 = 1 \cdot 3 \cdot 10, \text{ note that } 1 + 3 + 10 = 14.$$

$$30 = 1 \cdot 6 \cdot 5, \text{ note that } 1 + 6 + 5 = 12.$$

$$30 = 2 \cdot 3 \cdot 5, \text{ note that } 2 + 3 + 5 = 10.$$

We get that, the sum of these three different natural numbers can be equal to 10, 12, 14, 18. So, answer (D) cannot be equal to their sum. The correct answer is (D).

8. A natural number is called "successful" if its digits are consequent numbers in increasing order from left to right. For example, 45 and 567 are "successful" numbers. M is the smallest "successful" number greater than 100 and divisible by 67. What is the sum of the digits of M?

(A) 14   (B) 15   (C) 16   (D) 18   (E) 19

Answer. (A)

Solution. Here is the list of the first few "successful" numbers greater than 100.

$$123, 234, 345, 456, 567, 678, 789, 1234, 2345, \ldots$$

Note that the smallest number among these numbers that is divisible by 67 is 2345. The correct answer is (A).

9. The following paper shape can be cut only into rectangles. At least, how many cut-outs (rectangle pieces) can be there?

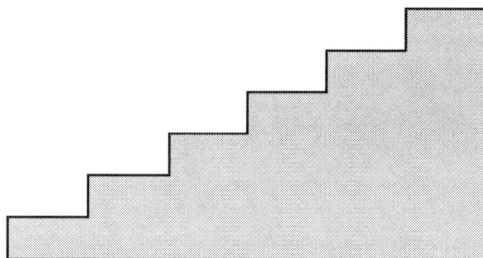

(A) 5   (B) 6   (C) 7   (D) 12   (E) 15

Answer. (B)

Solution. Consider the points A, B, C, D, E, F (see the picture). Note that any two of these points cannot belong to the same rectangle. So, there must be at least as many rectangles as the number of these points. As there are six points, then the overall number of rectangles cannot be less than six. In the figure below, an example for six rectangles is provided.

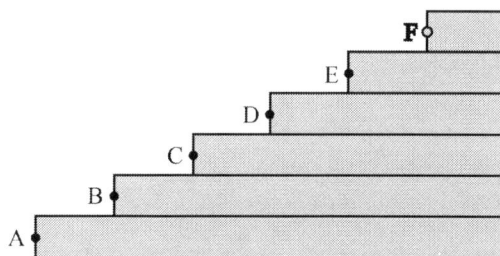

The correct answer is (B).

10. What is the number of all four-digit numbers that can be written using the digits 1, 2, 3, 4, so that each of these digits is used exactly once and the first digit is not 1, the second digit is not 2, the third digit is not 3, and the fourth digit is not 4?

(A) 6   (B) 9   (C) 10   (D) 12   (E) 20

Answer. (B)
Solution. All possible options are listed below.

$$2143, 2341, 2413,$$

$$3142, 3412, 3421,$$

$$4123, 4312, 4321.$$

The correct answer is (B).

Part B: Each correct answer is worth 4 points

11. Ann has some amount of money. If she decides to buy 11 copies of the same book, then she will have 50 cents left. If she decides to buy 15 such books, then she will need 70 cents more than she has. At least, how much more money (in cents) does Ann need to be able to buy 20 such books?

(A) 600   (B) 500   (C) 220   (D) 200   (E) 180

Answer. (C)

Solution. Let us denote the price of one such book by x cents.

Given that from one hand Ann has $11 \cdot x + 50$ cents and from the other hand she has $15 \cdot x - 70$ cents.

This means

$$11x + 50 = 15x - 70.$$

We get that

$$x = 30.$$

So, Ann has $11x + 50 = 11 \cdot 30 + 50 = 380$ cents.

Also, 20 such books cost $20 \cdot 30 = 600$ cents.

So, Ann needs at least $600 - 380 = 220$ cents to buy 20 such books. The correct answer is (C).

---

12. A square paper was folded into half twice, then a small part of it was cut out.

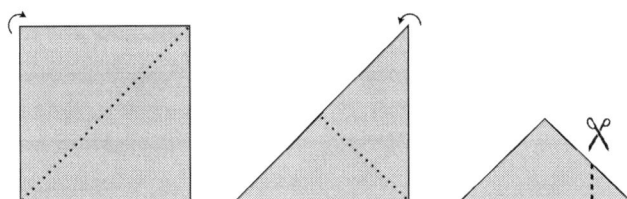

If we unfold it, which of the following shapes can we get?

    (A)   (B)   (C)   (D)   (E)

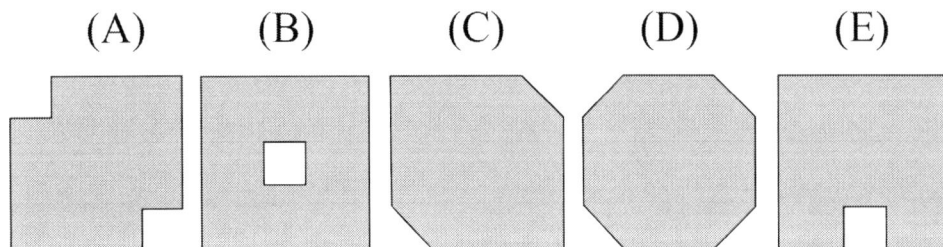

Answer. (A)

Solution. Note that the number of pieces is three and two of them must be squares. The correct answer is (A).

---

13. The height of each shelf of a kitchen rack is 25.5 cm. If we put four plates on top of each other their height is 6 cm. If we put seven such plates on top of each other their height is 9 cm.

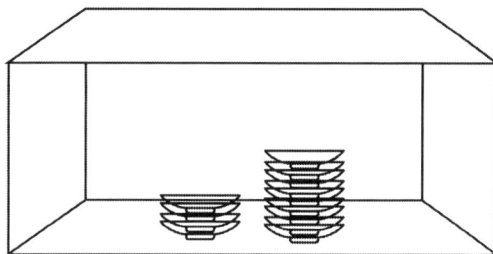

At most, how many plates can we put on top of each other in each shelf of this kitchen rack?

(A) 22   (B) 23   (C) 24   (D) 25   (E) 26

Answer. (B)

Solution. Given that if we put four plates on top of each other their height is 6 cm, and if we put seven such plates on top of each other their height is 9 cm. This means if we put three such plates (as 7 plates - 4 plates = 3 plates) on top of each other their height is 3 cm (as 9 cm - 6 cm = 3 cm). So, each plate increases the total height by 1 cm.

As it is given that if we put four plates on top of each other their height is 6 cm, then to find the height of the first plate from the total height 6 cm we need to subtract 3 cm (the height of three plates when we put these three plates on top of the first plate, that is $3 \cdot 1$ cm = 3 cm). So, the height of the first plate is 3 cm.

As the height of each shelf is 25.5 cm, the height of the first plate is 3 cm and each additional plate will increase the total height by 1 cm, then in each shelf on top of the first plate we can put at most 22 other plates (as $3 + 22 = 25 < 25.5$). So, in each shelf we can put at most 23 plates. The correct answer is (B).

14. How many times does the number 5 appear from 1 to 200?

(A) 35   (B) 38   (C) 39   (D) 40   (E) 41

Answer. (D)

Solution. The number 5 appears 11 times as the first digit: 5, 50, 51, 52, 53, 54, 55, 56, 57, 58, 59.

The number 5 appears 10 times as the middle digit: 150, 151, 152, 153, 154, 155, 156, 157, 158, 159.

The number 5 appears 19 times as the last digit: 15, 25, 35, 45, 55, 65, 75, 85, 95, 105, 115, 125, 135, 145, 155, 165, 175, 185, 195.

So, altogether the number 5 appears $11 + 10 + 19 = 40$ times. The correct answer is (D).

15. Given one large and four small rectangles (see the picture). Points A, B, C, D are the intersection points of the diagonals of each small rectangle. The area of the large rectangle is 20 and the area of each small rectangle is 2. What is the area of the shaded shape?

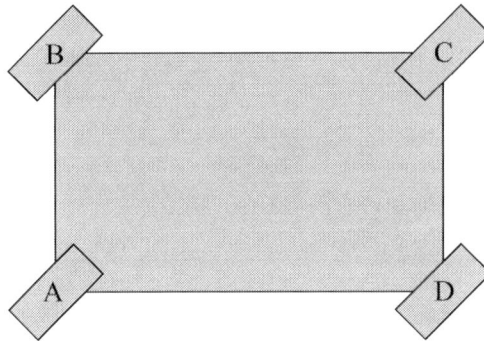

(A) 20   (B) 22   (C) 24   (D) 26   (E) 28

Answer. (D)

Solution. If we put together shapes a, b, c, d (see the figure), we get a small rectangle.

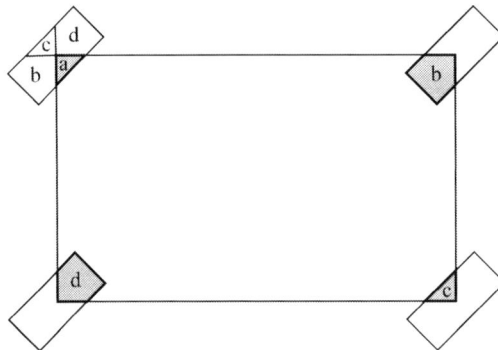

This means, the sum of the areas of shapes a, b, c, d is equal to 2. So, the area of the shaded shape is equal to the area of the large rectangle plus the area of all four small rectangles minus the area of shapes a, b, c, d, that is

$$20 + 4 \cdot 2 - 2 = 26.$$

The correct answer is (D).

16. A picket fence consists of 60 wooden pickets (see the picture). Each wooden picket needs to be colored either in red, blue, or orange, so that any three consecutive pickets have different colors. In how many different ways is it possible to color this picket fence?

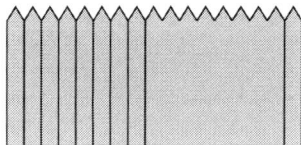

(A) 3^{60} (B) 81   (C) 27   (D) 18   (E) 6

Answer. (E)

Solution. The first three pickets can be colored in these 6 different ways:

$$red, blue, orange,$$

$$red, orange, blue,$$

$$blue, red, orange,$$

$$blue, orange, red,$$

$$orange, red, blue,$$

$$orange, blue, red.$$

Let us prove that when the first three pickets are colored, then starting from the fourth picket all the other pickets can be colored only in one way. For example, if the first three pickets are colored like this:

$$red, blue, orange,$$

let us show that the next picket (the fourth picket) can be only red.
Orange: The fourth picket cannot be orange, as we get this

$$red, blue, orange, orange, \ldots$$

This coloring does not work, as $blue, orange, orange$ are not different colors.
Blue: The fourth picket cannot be blue, as we get this

$$red, blue, orange, blue, \ldots$$

This coloring does not work, as $blue, orange, blue$ are not different colors.
Red: The fourth picket can be red, as we get this

$$red, blue, orange, red, \ldots$$

This coloring works, as any three consecutive pickets have different colors.

So, we can color the first three pickets in 6 different ways and all other pickets only in one way. The correct answer is (E).

17. In how many different ways can you draw a line from the top leftmost letter "k" to the bottom rightmost letter "k" to get the word "kayak"?

k	a	y
a	y	a
y	a	k

(A) 9   (B) 12   (C) 10   (D) 6   (E) 8

Answer. (D)
Solution. All 6 ways are shown below.

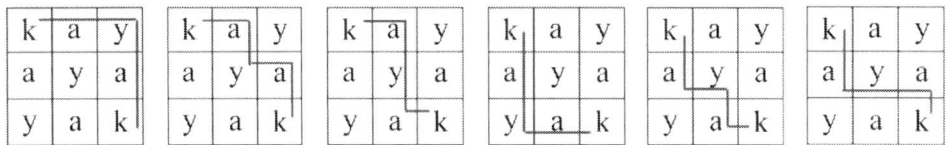

18. This shape consists of unit cubes (see the picture). How many unit squares make up the surface of this shape?

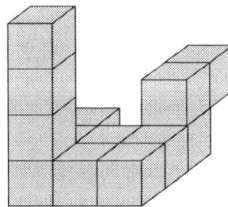

(A) 50   (B) 51   (C) 52   (D) 53   (E) 54

Answer. (A)
Solution. Altogether there are 13 cubes. Let us enumerate the cubes from 1 to 13.

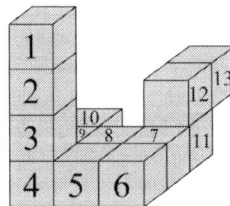

cube 1 (there are 5 surface squares),   cube 2 (there are 4 surface squares)
cube 3 (there are 4 surface squares),   cube 4 (there are 3 surface squares)
cube 5 (there are 3 surface squares),   cube 6 (there are 4 surface squares)
cube 7 (there are 3 surface squares),   cube 8 (there are 3 surface squares)
cube 9 (there are 3 surface squares),   cube 10 (there are 5 surface squares)
cube 11 (there are 4 surface squares),   cube 12 (there are 4 surface squares)
cube 13 (there are 5 surface squares). So, altogether there are

$$5 + 4 + 4 + 3 + 3 + 4 + 3 + 3 + 3 + 5 + 4 + 4 + 5 = 50$$

surface squares. The correct answer is (A).

19. 1, 1, 1, 1, 2, 2, 2, 2, 3 are the side lengths of three triangles. What is the product of the perimeters of these three triangles?

(A) 105　　　　(B) 120　　　　(C) 140　　　　(D) 150　　　　(E) 180

Answer. (A)

Solution. The only triangle that can have 3 as its side length is the triangle with side lengths 2, 2, 3.

The only triangle that can have 2 as its side length is the triangle with side lengths 1, 2, 2.

So, the side lengths of the third triangle are 1, 1, 1.

Then, the perimeter of the first triangle is 7, the perimeter of the second triangle is 5, and the perimeter of the third triangle is 3. So, the product of the perimeters of these three triangles is

$$7 \cdot 5 \cdot 3 = 105.$$

The correct answer is (A).

20. A floor has a shape of a regular hexagon of side length 10 meters (see the picture). At least, how many equilateral triangle tiles of side length 1 decimeter are needed to tile this floor?

(A) 400　　　(B) 600　　　(C) 6000　　　(D) 10000　　　(E) 60000

Answer. (E)

Solution. Using six equilateral triangle tiles of side length 1 decimeter we can get a regular hexagon of side length 1 decimeter (see the picture). So, the side length (1 decimeter) of this hexagon is 100 times smaller than the side length of the entire floor (10 meters). Then, the area of this regular hexagon is $100 \cdot 100$ times smaller than the area of the entire floor.

We know that 6 tiles are needed to tile this regular hexagon (of side length 1 decimeter), then at least $6 \cdot 100 \cdot 100 = 60000$ tiles are needed to tile the entire floor. The correct answer is (E).

Part C: Each correct answer is worth 5 points

21. A three-digit number is called an "interesting" number, if the sum of its digits is a prime number and the sum of any two digits is a prime number. How many three-digit "interesting" numbers are there?

(A) 8   (B) 9   (C) 13   (D) 17   (E) 18

Answer. (E)

Solution. The sum of any two odd digits is a prime number only if these two odd digits are 1 and 1. The sum of any two even digits is a prime number only if these two even digits are 2 and 0. So, three-digit "interesting" numbers must include either digits 1, 1 or 2, 0. We list all possible options:

$$111,$$
$$112, 121, 211,$$
$$114, 141, 411,$$
$$116, 161, 611,$$
$$203, 230, 302, 320$$
$$205, 250, 502, 520.$$

So, the total number of all three-digit "interesting" numbers is

$$1 + 3 + 3 + 3 + 4 + 4 = 18.$$

The correct answer is (E).

22. The airplane travels the distance between two cities in 3 hours 20 minutes. If the airplane increases its speed by 200 kilometers per hour, then it travels the same distance in 2 hours 30 minutes. What is the distance (in kilometers) between these two cities?

(A) 500   (B) 600   (C) 1000   (D) 2000   (E) 2100

110

Answer. (D)

Solution. Let the initial speed of the airplane be x kilometers per hour. As the airplane travels the distance between these two cities in 3 hours 20 minutes (which is the same as $3\frac{1}{3}$ hours), then the distance between these two cities is equal to $3\frac{1}{3} \cdot x$ kilometers.

Given also that if the airplane increases its speed by 200 kilometers per hour (so the new speed will be $x + 200$ kilometers per hour), then it travels the same distance in 2 hours 30 minutes (which is the same as 2.5 hours). So, the distance between these two cities is equal to $2.5 \cdot (x + 200)$. Then, we get that

$$3\frac{1}{3} \cdot x = 2.5 \cdot (x + 200).$$

That is the same as

$$\frac{10}{3} \cdot x = 2.5 \cdot x + 2.5 \cdot 200 = 2.5 \cdot x + 500.$$

We get that

$$10x = 3 \cdot (2.5x + 500).$$

$$10x = 7.5x + 1500.$$

$$10x - 7.5x = 1500.$$

$$2.5x = 1500.$$

$$x = 600.$$

So, the distance between these two cities is:

$$3\frac{1}{3} \cdot x = \frac{10}{3} \cdot 600 = 2000.$$

The correct answer is (D).

---

23. What is the sum of the digits of the smallest five-digit number so that none of its digits is divisible by each other?

(A) 30   (B) 31   (C) 32   (D) 33   (E) 34

Answer. (B)

Solution. Note that, if this five-digit number has two equal digits, then one of these two digits will be divisible by the other one. So, it cannot have two equal digits and all five digits of this five-digit number must be different. Note also that, none of its digits can be equal to 0 or 1. Otherwise, one of the digits will be divisible by the other one.

Let us consider the following two groups.

$\{2, 4, 8\}, \{3, 6\}, \{5\}, \{7\}, \{9\},$

$\{2, 4, 8\}, \{3, 9\}, \{5\}, \{6\}, \{7\}.$

Note that 5, 6, 7, 9 must be four of these five digits, otherwise however we choose any four digits from any of these two groups one of the digits will be divisible by the other one. Then, the fifth digit is either 4 or 8.

So, the smallest such five-digit number is 45679.

The correct answer is (B).

24. How many numbers with different digits are there that are greater than 2023 and smaller than 2320?

(A) 110   (B) 119   (C) 200   (D) 204   (E) 225

Answer. (B)

Solution. Any natural number that is greater than 2023 and smaller than 2320 is a four-digit number that starts with the digit 2. So, we want to find the number of four-digit numbers $2 \bigstar \heartsuit \clubsuit$, where $\bigstar$, $\heartsuit$, $\clubsuit$ are three different digits and $\bigstar$ can be either 0, 1, or 3 (as $2 \bigstar \heartsuit \clubsuit$ must be greater than 2023 and smaller than 2320, $\bigstar$ cannot be 2 because all digits of $2 \bigstar \heartsuit \clubsuit$ must be different and the digit 2 is already used).

So, let us consider the following cases.

If $\bigstar = 0$, we get $20 \heartsuit \clubsuit$. In this case, there are seven possible options to choose the digit $\heartsuit$ and $20 \heartsuit \clubsuit$ must be greater than 2023, so $\heartsuit$ can be 3, 4, 5, 6, 7, 8, 9. There are seven possible options to choose the digit $\clubsuit$ (as three digits will be already used 2, 0, and the digit $\heartsuit$). So, in this case the number of all such four-digit numbers $2 \bigstar \heartsuit \clubsuit$ is $7 \cdot 7 = 49$.

If $\bigstar = 1$, we get $21 \heartsuit \clubsuit$. In this case, there are eight possible options to choose the digit $\clubsuit$ (as two digits 2 and 1 are already used) and there are seven possible options to choose the digit $\heartsuit$ (as three digits will be already used 2, 1, and the digit $\clubsuit$). So, in this case the number of all such four-digit numbers $2 \bigstar \heartsuit \clubsuit$ is $8 \cdot 7 = 56$.

If $\bigstar = 3$, we get $23 \heartsuit \clubsuit$. In this case, the digit $\heartsuit$ can be either 0 or 1, because $23 \heartsuit \clubsuit$ must be smaller than 2320. There are seven possible options to choose the digit $\clubsuit$ (as three digits will be already used 2, 3, and the digit $\heartsuit$). So, in this case the number of all such four-digit numbers $2 \bigstar \heartsuit \clubsuit$ is $2 \cdot 7 = 14$.

So, there are $49 + 56 + 14 = 119$ numbers with different digits that are greater than 2023 and smaller than 2320.

---

25. What is the value of the following expression?

$$1011 - 1213 + 1415 - 1516 + 1617 - 1718 + 1819 - 2021 + \ldots + -9697 + 9899.$$

(A) 5555   (B) 5554   (C) 5553   (D) 5455   (E) 555

Answer. (D)
Solution. Note that
$$1011 = 100 \cdot 10 + 11,$$
$$1213 = 100 \cdot 12 + 13,$$
$$...$$
$$9697 = 100 \cdot 96 + 97,$$
$$9899 = 100 \cdot 98 + 99.$$

In general, for any four-digit number $\overline{abcd}$, we have

$$\overline{abcd} = 100 \cdot \overline{ab} + \overline{cd}.$$

Then, we get that

$$1011 - 1213 + 1415 - 1516 + 1617 - 1718 + 1819 - 2021 + ... + -9697 + 9899 =$$

$$= 100 \cdot (10 - 12 + 14 - 16 + ... - 96 + 98) + (11 - 13 + 15 - 17 - ... - 97 + 99) =$$

$$= 100(22 \cdot (-2) + 98) + 22 \cdot (-2) + 99 = 100 \cdot 54 + 55 = 5455.$$

26. What is the number of all two-digit numbers $\overline{ab}$, so that $\overline{ba}$ is also a two-digit number and when $\overline{ab}$ is divided by $\overline{ba}$ the remainder is $a + b$?

(A) 1   (B) 2   (C) 3   (D) 5   (E) 7

Answer. (C)
Solution. Given that when $\overline{ab}$ is divided by $\overline{ba}$ the remainder is $a + b$, this means the following

$$\overline{ab} = \overline{ba} \cdot q + a + b,$$

where q is a natural number. We get that

$$10a + b = (10b + a)q + a + b.$$

$$9a = 10bq + aq.$$

$$9a - aq = 10bq.$$

$$a(9 - q) = 10bq.$$

Note that $10bq$ is divisible by 5 and as $a(9-q) = 10bq$, then this means either $a = 5$ or $9 - q = 5$. If $a = 5$, we get $9 = q(2b + 1)$. So, either $b = 1$ or $b = 4$. In this case, we get the following two-digit numbers: 51, 54.
If $9 - q = 5$, we get $q = 4$ and $a = 8b$. So, either $b = 1$ or $a = 8$. In this case, we get the following two-digit number: 81.
So, we get the following two-digit numbers: 51, 54, 81.
The correct answer is (C).

113

27. There are 17 tables in a restaurant. The following placements are possible: tables that are placed separately, tables that are placed in pairs, tables that are placed in triplets. There are four chairs next to each table placed separately, there are six chairs next to each table placed in pairs, and there are eight chairs next to each table placed in triplets. Given that altogether there are 50 chairs. What is the difference of the number of tables placed in triplets and the number of tables placed separately?

(A) 1 (B) 2 (C) 3 (D) 4 (E) 5

Answer. (A)

Solution. Let x be the number of tables that are placed separately. Let y be the number of tables that are placed in pairs, and let z be the number of tables that are placed in triplets. Given that there are 17 tables in the restaurant, so we get

$$x + 2y + 3z = 17.$$

Given also that there are 50 chairs in the restaurant, so we get

$$4x + 6y + 8z = 50.$$

Multiplying both sides of the first equation by 3, we get

$$3(x + 2y + 3z) = 3 \cdot 17 = 51.$$

We have

$$3x + 6y + 9z = 51,$$

and

$$4x + 6y + 8z = 50.$$

Subtracting the last equation from the one before it, we get

$$3x + 6y + 9z - (4x + 6y + 8z) = 51 - 50 = 1.$$

This means

$$z - x = 1.$$

So, the difference of the number of tables placed in triplets and the number of tables placed separately is equal to 1. The correct answer is (A).

---

28. At most, how many such shapes (in any position: rotated or flipped) can be cut out from a 7×9 paper rectangle? Each of these shapes must consist of four unit squares.

(A) 10   (B) 11   (C) 12   (D) 13   (E) 14

Answer. (C)

Solution. Note that each such shape (that consists of four unit squares) includes exactly one colored square (see the picture).

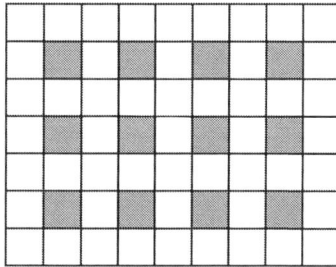

So, the number of cut out shapes is not more than 12 (as there are 12 colored squares). The example below shows how can 12 such shapes be cut out from a 7×9 paper rectangle.

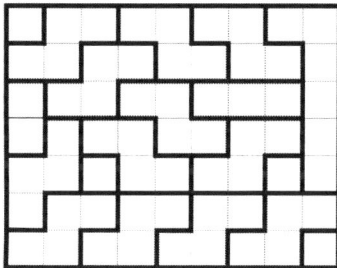

The correct answer is (C).

29. A 4×4 square consists of 16 unit squares. At least how many sides of these 16 squares must we erase, so that each square consisting of the sides of these unit squares includes at least one erased side?

(A) 6   (B) 9   (C) 10   (D) 5   (E) 8

Answer. (B)

Solution. Let n sides of unit squares be erased and the condition of the problem holds true. Let us color this 4×4 square as a chessboard (see the picture).

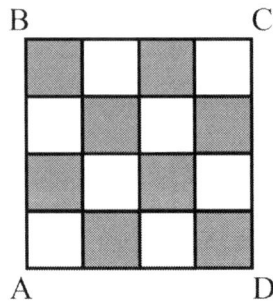

According to our assumption, from one of the sides AB, BC, CD, DA a side of a unit square was erased. Without loss of generality, we can assume that it is a side of a black unit square. Now, let us consider 8 white unit squares. From any of these 8 white unit squares at least one side was erased, so

$$n \geq 1 + 8 = 9.$$

Below we provide an example for $n = 9$.

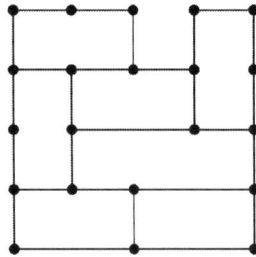

We get that at least 9 unit squares need to be erased.
The correct answer is (B).

---

30. The numbers 1, 2,..., 9 are written in the squares of a 3×3 square (one number per square). Consider all three row sums and all three column sums. At most, how many of these six numbers can be prime numbers?

(A) 2   (B) 3   (C) 4   (D) 5   (E) 6

Answer. (C)

Solution. Note that if any row (column) sum is a prime number, then that number must be 7, 11, 13, 17, 19, 23. The sum of all numbers from 1 to 9 is $1 + 2 + ... + 9 = 45$ and 45 can be written as the sum of three prime numbers in one of the following ways:

$$45 = 7 + 19 + 19,$$

$$45 = 11 + 17 + 17,$$

$$45 = 13 + 13 + 19.$$

Then, three row (columns) sums cannot be different prime numbers (at least two of them are the same). So, the answer is not greater than 4. Below we provide an example where the number of different primes is 4.

4	1	2	7
6	5	8	19
7	9	3	
17		13	

The correct answer is (C).

Solutions of Test 5

1. To make the equation $2050 + \bigstar - 1 = 2051$ correct, by which of the following must we replace $\bigstar$?

(A) 0   (B) 1   (C) 2   (D) 2050   (E) 2051

Answer. (C)
Solution. If we replace $\bigstar$ by 0, then $2050 + 0 - 1 = 2049$.
If we replace $\bigstar$ by 1, then $2050 + 1 - 1 = 2050$.
If we replace $\bigstar$ by 2, then $2050 + 2 - 1 = 2052 - 1 = 2051$.
If we replace $\bigstar$ by 2050, then $2050 + 2050 - 1 = 4100 - 1 = 4099$.
If we replace $\bigstar$ by 2051, then $2050 + 2051 - 1 = 4101 - 1 = 4100$.
So, $\bigstar$ must be replaced by 2 to get 2051. The correct answer is (C).

2. The sum of all six numbers written at the vertices of each hexagon is equal to 15 (see the picture).

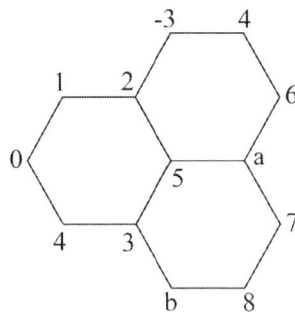

What is the value of $a - b$?

(A) 4   (B) 6   (C) 8   (D) 10   (E) 12

Answer. (D)

Solution. Given that the sum of all six numbers written at the vertices of each hexagon is equal to 15, then

$$2 + 5 + (-3) + 4 + 6 + a = 15.$$

This means

$$14 + a = 15.$$

So, we get

$$a = 1.$$

As the sum of all six numbers written at the vertices of each hexagon is equal to 15, then

$$3 + 5 + a + 7 + 8 + b = 15.$$

As $a = 1$, we get

$$3 + 5 + 1 + 7 + 8 + b = 15.$$

This means

$$24 + b = 15.$$

So, we get

$$b = -9.$$

As $a = 1$ and $b = -9$, then

$$a - b = 1 - (-9) = 1 + 9 = 10.$$

The correct answer is (D).

---

3. What is the sum of all whole numbers in between -12.01 and 14.03?

(A) 14   (B) 27   (C) 30   (D) 31   (E) 32

Answer. (B)

Solution. All whole numbers in between -12.01 and 14.03 are

$$-12, -11, -10, -9, -8, -7, -6, -5, -4, -3, -2, -1,$$

$$0, 1, 2, 3, 4, 5, 6, 7, 8, 9, 10, 11, 12, 13, 14.$$

We have

$$-12 + 12 = 0, \quad -11 + 11 = 0, \quad -11 + 11 = 0, \quad -10 + 10 = 0,$$

$$-9 + 9 = 0, \quad -8 + 8 = 0, \quad -7 + 7 = 0, \quad -6 + 6 = 0,$$

$$-5 + 5 = 0, \quad -4 + 4 = 0, \quad -3 + 3 = 0, \quad -2 + 2 = 0, \quad -1 + 1 = 0.$$

This means, the sum of all whole numbers in between -12.01 and 14.03 is equal to $13 + 14 = 27$.
The correct answer is (B).

4. How many of the numbers 0, 1, 2, 3, 4, 5, 6, 7, 8, 9 have an *axis of symmetry?*

(A) 2   (B) 1   (C) 5   (D) 3   (E) 4

> **Answer. (D)**
> **Solution.** From the list of the numbers 0, 1, 2, 3, 4, 5, 6, 7, 8, 9 only the numbers 0, 3, and 8 have an axis of symmetry (see the picture).
>
> $$0 \quad 0 \quad 3 \quad 8 \quad 8$$
>
> The correct answer is (D).

5. Given that

$$\frac{5}{6} + \frac{7}{50} = \frac{m}{300}.$$

What is the sum of the digits of m?

(A) 10   (B) 12   (C) 13   (D) 15   (E) 17

> **Answer. (C)**
> **Solution.** We have
>
> $$\frac{5}{6} + \frac{7}{50} = \frac{125 + 21}{150} = \frac{146}{150} = \frac{292}{300}.$$
>
> We get
>
> $$\frac{292}{300} = \frac{m}{300}.$$
>
> This means
>
> $$m = 292.$$
>
> So, the sum of the digits of m is $2 + 9 + 2 = 13$. The correct answer is (C).

6. There are some red, blue, and black pens on a table. Exactly two of them are not black and exactly three of them are not red. How many black pens are there on the table?

(A) 1   (B) 2   (C) 3   (D) 4   (E) 5

> **Answer. (B)**
> **Solution.** Given that exactly two are not black, this means there must be one red and one blue pen on the table. Given also that exactly three of the are not red and we know that there is one blue pen, this means there are 2 black pens. The correct answer is (B).

7. A paper square was cut into one square of size 2×2 and n rectangles of size 1×3, where n is a natural number. What is the smallest possible value of n?

(A) 2   (B) 3   (C) 4   (D) 5   (E) 7

Answer. (C)

Solution. If m is the side length of the paper square, then the paper square consists of $m \cdot m = m^2$ squares of size 1×1. Given that a paper square was cut into one square of size 2×2 and n rectangles of size 1×3, where n is a natural number. This means, the paper square consists of $4 + 3 \cdot n$ squares of size 1×1. So, we get

$$m^2 = 4 + 3 \cdot n,$$

where m is a natural number. Then, to solve the problem, we need to find the smallest possible n so that $4 + 3 \cdot n$ is a perfect square. When $n = 1$, $n = 2$, or $n = 3$, then $4 + 3 \cdot n$ is not a perfect square. When $n = 4$, then $4 + 3 \cdot n = 4 + 3 \cdot 4 = 16 = 4^2$. We provide an example how 4×4 paper square can be cut into one square of size 2×2 and 4 rectangles of size 1×3.

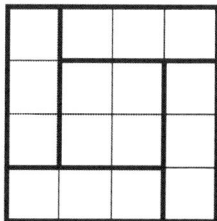

The correct answer is (C).

8. At most how many different digits must be used in order to write three consecutive three-digit numbers?

(A) 5   (B) 6   (C) 7   (D) 8   (E) 9

Answer. (B)

Solution. For example three consecutive three-digit numbers 239, 240, 241 can be written using six different digits, namely the digits 0, 1, 2, 3, 4, 9. Using seven or more different digits is not possible, because the first two digits of at least two of these numbers must be the same (for example in the case of 239, 240, 241 the last two numbers start with 2 and 4). Moreover, either the first digit of the other number is the same also (for example in the case of 239 the digit 2 is the same as for 240 and 241) or its last two digits are 99 (for example 399, 400, 401). In both cases, we get that seven or more different digits is not possible. So, at most six different digits must be used. The correct answer is (C).

9. A digital clock is broken and does not show : symbol. So, instead of showing 17 : 30 it shows 1730. At most, how many different four-digit numbers can it show in one day?

(A) 840   (B) 841   (C) 960   (D) 1200   (E) 1440

> **Answer. (A)**
> **Solution.** If it is $A : B$ o'clock, then the clock shows AB. To find the number of all different four-digit numbers, A must be a two-digit number and B must be a two-digit number. So, A can be 10, 11,..., 23 and B can be 00, 01,..., 59. There are 14 numbers from 10 to 23 and there are 60 numbers from 00 to 59. So, there are $14 \cdot 60 = 840$ different possibilities. The correct answer is (A).

10. Eight identical rectangles form a large rectangle (see the figure). What is the ratio of the length to the width of the large rectangle?

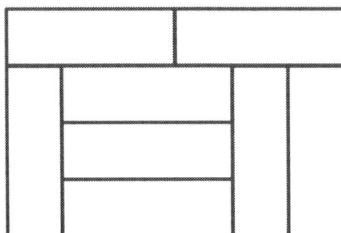

(A) 3:1   (B) 4:1   (C) 4:3   (D) 3:2   (E) 5:3

> **Answer. (D)**
> **Solution.** Let the width of the smaller rectangle be 1 unit, then from the figure we get that its length must be three times larger (so 3 units). We also get (from the figure) that the width of the larger rectangle is $1 + 3 = 4$ units and its length is $3 + 3 \cdot 1 = 6$ units. So, the ratio of its length to its width is $6 : 4 = 3 : 2$. The correct answer is (D).

Part B: Each correct answer is worth 4 points

11. A chocolate bar consists of 32 squares. One almond is placed on any two squares that have a common side, so that half of the almond is in one square and the other half is in the other square (see the picture). How many almonds are used?

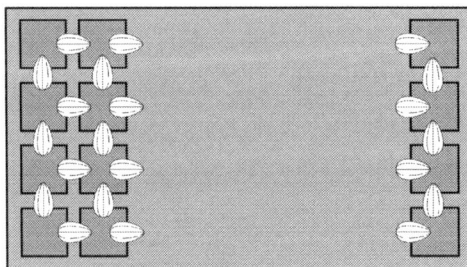

(A) 21   (B) 24   (C) 40   (D) 50   (E) 52

12. 12 chairs are placed next to a round table. At least two people sit on these chairs, each chair can be used only by one person. For any two people, at least one of them has two neighbors (is sitting in between of these two neighbors). How many people are there?

(A) 6   (B) 8   (C) 6 or 8   (D) cannot be determined   (E) other answer

Answer. (E)
Solution. If there is a chair that no one sits on it, then consider two neighbor chairs of that chair (the left neighbor chair and the right neighbor chair). So, either no one sits on any of these two chairs, or at least one of these two chairs is being used. Then, consider two neighbor chairs of each of these chairs. If we continue like this, then sooner or later we will reach to two chairs that no one sits on them but someone sits on one of their neighbor chairs. This is not possible, as given that for any two people, at least one of them has two neighbors.
This means, there is no chair that no one sits on it. So, all chairs are being used. Then, there are 12 people. The correct answer is (E).

13. At least, by how many straight lines must this paper shape be cut so that putting the pieces together we can make a hexagon?

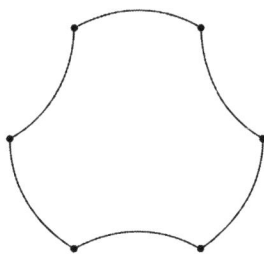

(A) 1   (B) 2   (C) 3   (D) 4   (E) 6

Answer. (C)

Solution. The picture below shows an example how this paper shape can be cut by 3 straight lines, so that putting the pieces together we can make a hexagon.

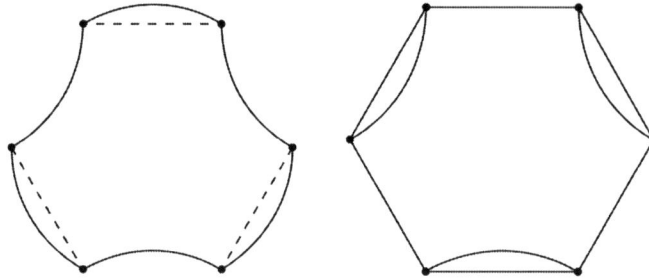

If we cut this shape by 1 or 2 straight lines and put the pieces together, then the border of the new shape cannot have more than 4 straight lines and we will not be able to make a hexagon. So, we must use at least 3 straight lines. The correct answer is (C).

14. A big rectangle is divided into five rectangles (see the picture). The sum of the perimeters of four corner rectangles is 2024. What is the perimeter of the big rectangle?

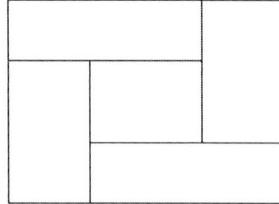

(A) 2024　　　　(B) 1012　　　　(C) 1000　　　　(D) 4048　　　　(E) 2000

Answer. (B)

Solution. The perimeter of the big rectangle is the double of the length of the bold part. So, the sum of the perimeters of four corner rectangles is the double of four such bold parts. Given that the sum of the perimeters of four corner rectangles is 2024. Then, the sum of the lengths of these four bold parts is the half of 2024. So, the perimeter of the big rectangle is $2024 : 2 = 1012$. The correct answer is (B).

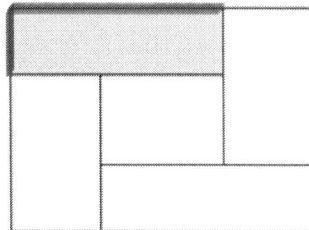

124

15. What is the sum of the digits of the smallest natural number that the product of its digits is equal to 648?

(A) 12   (B) 15   (C) 16   (D) 25   (E) 26

> Answer. (E)
> **Solution.** The product of the digits of a two-digit number can be at most $9 \times 9 = 81$. So, it cannot be equal to 648. Can it be a three-digit number? We have $648 = 8 \times 9 \times 9$. So, one of its digits must be 8 and the other two digits must be 9.
> The smallest such number is 899.
> The correct answer is (E).

16. The sum of three different natural numbers is 10. Which of the following numbers cannot be equal to their product?

(A) 14   (B) 16   (C) 18   (D) 20   (E) 30

> Answer. (B)
> **Solution.** We have
> $$10 = 1 + 2 + 7,$$
> $$10 = 1 + 3 + 6,$$
> $$10 = 1 + 4 + 5,$$
> $$10 = 2 + 3 + 5,$$
>
> So, we get
> $$1 \cdot 2 \cdot 7 = 14,$$
> $$1 \cdot 3 \cdot 6 = 18,$$
> $$1 \cdot 4 \cdot 5 = 20,$$
> $$2 \cdot 3 \cdot 5 = 30.$$
>
> So, their product can be 14, 18, 20, and 30.
> The correct answer is (B).

17. In how many different ways is it possible to connect the letters to form the word AMERICA (see the picture)? Each letter can be connected only with its neighbor letters.

A	M	E	R
M	E	R	I
E	R	I	C
R	I	C	A

(A) 10   (B) 20   (C) 21   (D) 25   (E) 50

Answer. (B)

Solution. The leftmost letter A can be included in the word AMERICA only in one way and the letter E from the second row can be included in the word AMERICA in two different ways (see the picture).

A	M	E	R
M	E	R	I
E	R	I	C
R	I	C	A

In each square let us write the number that shows in how many different ways can that letter be included in the word AMERICA (see the picture).

1	1	1	1
1	2	3	4
1	3	6	10
1	4	10	20

So, there are 20 different ways to connect these letters to form the word AMERICA. The correct answer is (B).

18. 30 students sit in 3 rows. Half of the students from the third row moved to the first two rows and the number of students in each of the first two rows doubled. How many students were in the third row?

(A) 10   (B) 14   (C) 18   (D) 20   (E) 24

> Answer. (D)
>
> **Solution.** Given that half of the students from the third row moved to the first two rows and the number of students in each of the first two rows doubled, this means that in the third row there were twice more students than in the first two rows together. So, if the number of students in the first two rows was x, then the number of students in the third row was $2 \cdot x$. Altogether there are 30 students, so $x + 2 \cdot x = 30$. Then, $x = 10$. So, there were 10 students in the first two rows and there were 20 students in the third row. The correct answer is (D).

19. At most, how many such shapes ⌐ (in any position: rotated or flipped) is possible to cut out from a 5×5 paper square? Each shape must consist of three squares of size 1×1.

(A) 5   (B) 6   (C) 7   (D) 8   (E) 9

> Answer. (D)
>
> **Solution.** A 5×5 square consists of 25 squares of size 1×1 and each such shape ⌐ consists of three squares of size 1×1. So, we cannot have 9 or more such shapes, as $3 \cdot 9 > 25$.
> We provide an example, where it is possible to cut out 8 such shapes from a 5×5 paper square (see the picture).
>
>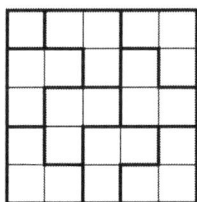
>
> The correct answer is (D).

20. In each square of a 3×3 square is written a number. At least how many different numbers must be written in this 3×3 square, so that all six row and column sums are different from each other? (Row sum is the sum of three numbers in that row, column sum is the sum of three numbers in that column. There are three row sums and three column sums, so six sums).

(A) 1   (B) 2   (C) 3   (D) 4   (E) 5

Answer. (C)

Solution. There must be more than two different numbers written in this 3×3 square. Clearly, if there is only one number written that all row and column sums will be the same. If there are only two different numbers written, then there can be only four possible values for each row and/or column sum $3a$, $2a + b$, $a + 2b$, $3b$. As we have only four different options, no matter how we choose six numbers from here at least two of them will be the same. So, there must be at least three different numbers written in this 3×3 square. We provide an example for three different numbers, for example

4	4	4	12
4	0	0	4
1	1	4	6
9	5	8	

The correct answer is (C).

21. An *Euler path* is a walk along the edges of this shape (of an edge length 1) which uses every edge exactly once. What is the length of the longest *Euler path* from A to B?

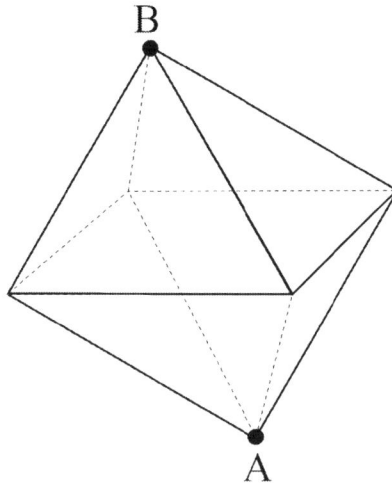

(A) 12   (B) 11   (C) 10   (D) 9   (E) 8

Answer. (C)
Solution. We start at A, so at least one of its four edges is not used (otherwise the path would lead back to A). We must end up in B, so at least one of its four edges is not used. Then, the length of the longest *Euler path* from A to B can be at most 10. Here is an example of *Euler path* of length 10.

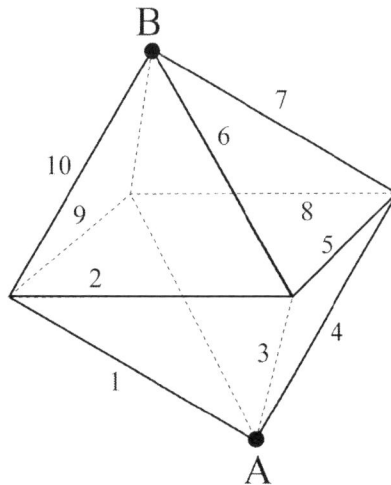

The correct answer is (C).

22. A natural number was written in each square of a 4×4 square. Some of these numbers were erased (see the picture). We know that the products of all four numbers of any row (column) are equal and can be written as a product of three prime numbers. What is a?

			a
2			
3	2		
5	3	2	

(A) 1   (B) 2   (C) 3   (D) 4   (E) 5

Answer. (B)

Solution. As the product of all four numbers of the first column (leftmost column) can be written as a product of three primes, then let us show that $b = 1$ (see the picture). If b is not 1, then in the first column there will be 5, 3, 2, and another natural number. So, there will be at least 4 primes.

b	c	e	a
2	d		
3	2		
5	3	2	

As the product of all four numbers of the second column must be equal to the product of all four numbers of the first column (which is 30), then we get $c \cdot d = 5$.

c cannot be 1, as $b = 1$ and is $c = 1$, then the top row will have two 1's and it will not have three primes. So, $c = 5$, $d = 1$. As the product of all four numbers of the top row must be 30, then $e \cdot a = 6$. e cannot be 2, as there is already 2 in the third column. So, we get $e = 3$ and $a = 2$. The correct answer is (B).

23. Winnie the pooh can eat one honey jar in 3 hours. Piglet can eat the same honey jar in 6 hours. In how many hours can they both together eat one honey jar?

(A) 2.5   (B) 2   (C) 1.5   (D) 1   (E) 0.5

Answer. (B)
Solution. Winnie eats one honey jar in 3 hours, so in one hour Winnie eats three times less. Piglet eats one jar in 6 hours, so in one hour Piglet eats six times less. In one hour Winnie and Piglet eat the sum of both. As the shaded part is half of the jar and both together they eat the shaded part in one hour, then to eat the whole jar they need twice more time (2 hours). The correct answer is (B).

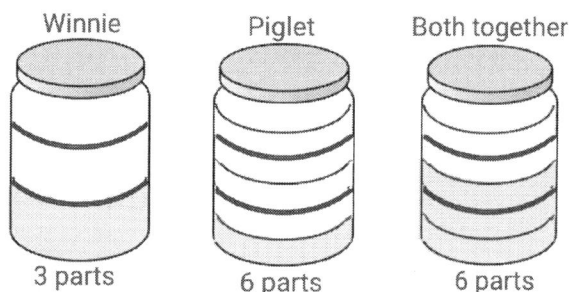

24. How many nine-digit numbers are there with the digits in decreasing order? For example, 976543210 is such a nine-digit number.

(A) 3   (B) 5   (C) 8   (D) 9   (E) 10

Answer. (E)
Solution. At first, we write the following ten-digit number 9876543210. We know that all nine-digit numbers with the digits in decreasing order were obtained by erasing one of the digits of this ten-digit number. For example, when we erase the digit 8, we get the number 976543210. As there are ten digits written in the number 9876543210 and we can erase any of this ten digits, then we will end up with ten different nine-digit numbers with the digits in decreasing order. The correct answer is (E).

25. Given a 3×3 square, where $a > 75 > b$. Consider all six row and column sums. Given that two of these sums are equal. What is $a + b$?

a	a	a
a	b	b
75	75	a

(A) 75   (B) 150   (C) 200   (D) 210   (E) 250

Answer. (B)

Solution. These six row and column sums are $3a$, $2a + 75$, $2a + b$, $a + 75 + b$, $a + 2b$, and $a + 150$. Note that

$$3a > 2a + 75 > 2a + b > a + 75 + b > a + 2b.$$

This means that none of these five sums can be equal to each other, so the sixth sum $a + 150$ must be equal to one of these sums. We also have that

$$a + 75 + b < a + 150 < 2a + 75.$$

This means that $a + 150$ can be equal only to $2a + b$.

$$a + 150 = 2a + b.$$

We get

$$a + b = 150.$$

The correct answer is (B).

---

26. The sum of two-digit numbers $\overline{ab}$, $\overline{bc}$, $\overline{ca}$ is a perfect square. What is the value of $a + b + c$?

(A) 9   (B) 10   (C) 11   (D) 12   (E) 24

Answer. (C)

Solution. We have

$$\overline{ab} + \overline{bc} + \overline{ca} = (10 \cdot a + b) + (10 \cdot b + c) + (10 \cdot c + a) =$$

$$= 11 \cdot a + 11 \cdot b + 11 \cdot c = 11 \cdot (a + b + c).$$

Given that $\overline{ab} + \overline{bc} + \overline{ca}$ is a perfect square, this means $11 \cdot (a + b + c)$ is a perfect square. As a, b, c are digits, then

$$3 \leq a + b + c \leq 27.$$

In order $11 \cdot (a + b + c)$ to be a perfect square $a + b + c$ must have the number 11 as one of its factors. As $3 \leq a + b + c \leq 27$ and as $a + b + c$ is divisible by 11, we get $a + b + c = 11$ or $a + b + c = 22$. If $a + b + c = 22$, then $11 \cdot 22$ is not a perfect square. If $a + b + c = 11$, then $11 \cdot 11$ is a perfect square. The correct answer is (C).

27. How many five-digit numbers are there with these two properties?
• Sum of its digits is 2. For example, 10001, as $1 + 0 + 0 + 0 + 1 = 2$.
• There is a natural number, so that when it is added to the five-digit number, then the sum of the digits of their sum is also 2. For example, for 10001 such natural number is 9, as $10001 + 9 = 10010$ and $1 + 0 + 0 + 1 + 0 = 2$.

(A) 0   (B) 1   (C) 2   (D) 3   (E) 4

Answer. (E)

Solution. There is no such five-digit number starting with 2, 3,..., 9, as only 20000 satisfies the first property ($2 + 0 + 0 + 0 + 0 = 2$), but it does not satisfy the second property (if we add any natural number to it, then the sum of the digits of their sum will be greater than 2). So, the first digit of such five-digit number must be 1. As the sum of all five digits must be 2, then it means that it must have another digit 1 and three digits 0. Such five-digit numbers are 10001, 10010, 10100, 11000. All for numbers satisfy both properties, as we have

$$10001 + 9 = 10010 \text{ and } 1 + 0 + 0 + 1 + 0 = 2,$$

$$10010 + 90 = 10100 \text{ and } 1 + 0 + 1 + 0 + 0 = 2,$$

$$10100 + 900 = 11000 \text{ and } 1 + 1 + 0 + 0 + 0 = 2,$$

$$11000 + 9000 = 20000 \text{ and } 2 + 0 + 0 + 0 + 0 = 2.$$

So, there are four such five-digit numbers. The correct answer is (E).

28. In how many different ways is it possible to choose three numbers from 1, 2,..., 30, so that their product is 2024 and they are in increasing order? For example $8 \cdot 11 \cdot 23 = 2024$.

(A) 1   (B) 4   (C) 3   (D) 2   (E) 5

Answer. (D)

Solution. We have
$$2024 = 2^3 \cdot 11 \cdot 23.$$

As 23 is prime, then one of the numbers must be 23. The other number must be divisible by 11 and be less than or equal to 30, so it can be either 11 or 22. If we choose 11 and 23, then the third number is 8. So, 8, 11, 23 are in increasing order and their product is 2024. If we choose 22 and 23, then the third number is 4. So, 4, 22, 23 are in increasing order and their product is 2024. So, we can choose in two different ways three numbers from 1, 2,..., 30, so that their product is 2024 and they are in increasing order. The correct answer is (D).

29. A 3×3 square consists of nine squares. At least how many sides of these nine squares must be removed, so that each square has at least one side that was removed?

(A) 3   (B) 4   (C) 5   (D) 6   (E) 7

Answer. (D)

Solution. Let us color this 3×3 square as a chessboard (see the figure).

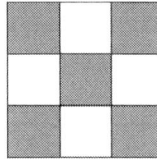

Let n be the number of removed squares. Each shaded square must have at least one removed side, so $n \geq 5$. Let us prove that $n = 5$ is not possible.

If $n = 5$, then exactly one side of each shaded square was removed. Let us mark the remaining sides (see the figure).

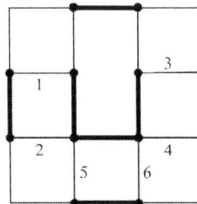

According to the condition of the problem, at least one side of each of pairs (1, 2), (3, 4), (5, 6) is removed. So, we have one of the following situations (see the figure).

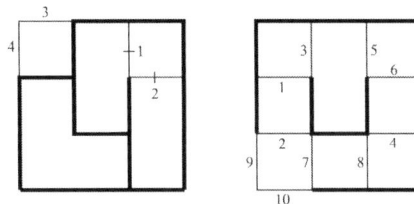

The left figure is not possible, as from each of pairs (1, 2) and (3, 4) one side is removed. The right figure is also not possible, as from each of pairs (1, 2), (3, 4), (5, 6), (7, 8), (9, 10) at least one side is removed. So, $n \geq 6$. We provide an example for $n = 6$ (see the figure).

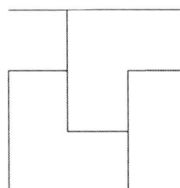

The correct answer is (D).

30. Using three different digits it is possible to write six different three-digit numbers. For example, if we use the digits 1, 2, 3, we can write 123, 132, 213, 231, 312, 321. If we use the digits 2, 3, 4, we can write 234, 243, 324, 342, 423, 432. Some three different digits were used to write six different three-digit numbers, so that the sum of five of them is 2026. What is the product of the digits of the sixth three-digit number?

(A) 20   (B) 21   (C) 24   (D) 30   (E) 36

Answer. (C)

Solution. Let a, b, c be different digits used to write the following six different three-digit numbers $\overline{abc}$, $\overline{acb}$, $\overline{bac}$, $\overline{bca}$, $\overline{cab}$, $\overline{cba}$. The sum of these six three-digit numbers is

$$100 \cdot (2a + 2b + 2c) + 10 \cdot (2a + 2b + 2c) + (2a + 2b + 2c) = 222 \cdot (a + b + c).$$

Given that the sum of five of these numbers is 2026, so

$$222 \cdot (a + b + c) = 2026 + \text{the sixth three-digit number}.$$

This means that

$$a + b + c = \frac{2026}{222} + \frac{\text{the sixth three-digit number}}{222}.$$

Note that $4.6 > \frac{\text{any three-digit number}}{222} > 0.4$ and that $9.2 > \frac{2026}{222} > 9.1$, so $13.8 > a + b + c > 9.5$. Then, $a + b + c$ can be 10, 11, 12, 13. So, we need to check which of these values satisfies

$$222 \cdot (a + b + c) = 2026 + \text{the sixth three-digit number}.$$

Only $a + b + c = 11$ satisfies and we get that the sixth three-digit number is 416. So, the product of its digits is $4 \cdot 1 \cdot 6 = 24$. The correct answer is (C).

Solutions of Test 6

1. Which of the following expressions is equal to 1?

(A) $1000 - \frac{997}{3}$ (B) $\frac{1000}{3} - 997$ (C) $\frac{1000}{3} - \frac{997}{3}$ (D) $997 - \frac{1000}{3}$ (E) $\frac{997}{3} - \frac{1000}{3}$

Answer. (C)

Solution. We have

$$(A) \qquad 1000 - \frac{997}{3} = 1000 - 332\frac{1}{3},$$

this is not a whole number and cannot be equal to 1.

$$(B) \qquad \frac{1000}{3} - 997 = 333\frac{1}{3} - 997,$$

this is not a whole number and cannot be equal to 1.

$$(C) \qquad \frac{1000}{3} - \frac{997}{3} = \frac{1000 - 997}{3} = \frac{3}{3} = 1.$$

$$(D) \qquad 997 - \frac{1000}{3} = 997 - 333\frac{1}{3},$$

this is not a whole number and cannot be equal to 1.

$$(E) \qquad \frac{997}{3} - \frac{1000}{3} = \frac{997 - 1000}{3} = \frac{-3}{3} = -1.$$

So, the correct answer is (C).

2. A wire was folded twice (see the picture). How long is this wire?

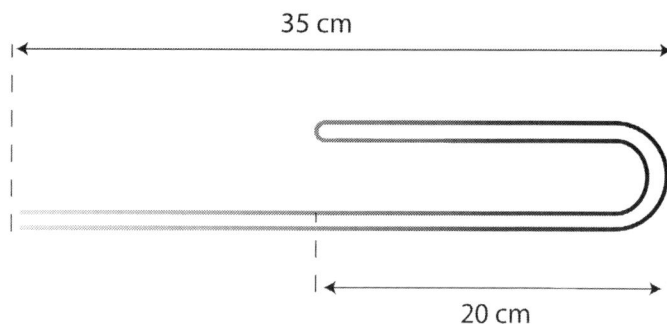

(A) 100 cm   (B) 110 cm   (C) 115 cm   (D) 120 cm   (E) 125 cm

Answer. (B)
Solution. As the wire was folded twice, then the length of the wire is:

$$2 \cdot 35 + 2 \cdot 20 = 110 \text{ cm.}$$

The correct answer is (B).

3. The average of two numbers is 10. If one of the numbers is 12, what is the other number?

(A) 10   (B) 6   (C) 9   (D) 12   (E) 8

Answer. (E)
Solution. Given that one of the numbers is 12. Let the other number be x, as their average is 10, we have

$$\frac{12 + x}{2} = 10.$$

$$12 + x = 2 \cdot 10 = 20.$$

So, we get

$$x = 8.$$

The correct answer is (E).

4. Given the points X(5), Y(4), Z(7) on the number line (see the picture). What is their correct order from left to right?

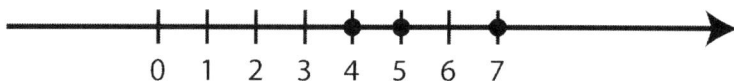

(A) X, Y, Z   (B) X, Z, Y   (C) Y, X, Z   (D) Y, Z, X   (E) Z, X, Y

5. Some Australian kangaroos can jump either 3 meters or 5 meters. At least, how many jumps are needed to travel a distance of exactly 48 meters?

(A) 16   (B) 12   (C) 11   (D) 9   (E) 10

Answer. (E)
Solution. Each jump is either 3 meters or 5 meters. We want to have as few jumps as possible, so we need to choose as many 5 meters jumps as possible. It is not possible to travel a distance of exactly 48 meters only by 5 meter jumps, as 48 is not divisible by 5. So, at least one 3 meters jump must be used. $48 - 3 = 45$ and 45 is divisible by 5 (we have $\frac{45}{5} = 9$). So, nine 5 meters jumps and one 3 meters jump can be used to travel a distance of exactly 48 meters. Altogether, there will be 10 jumps. The correct answer is (E).

6. Given two 1×3 paper rectangles, one black and one white. Which of the following shapes is not possible to get from these two paper rectangles? You cannot fold the rectangles.

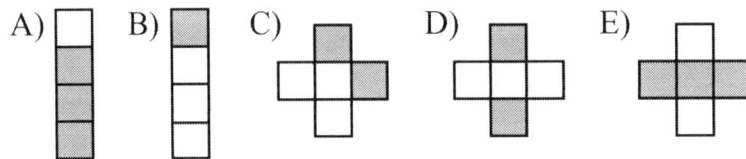

(A)   (B)   (C)   (D)   (E)

Answer. (C)
Solution. It is not possible to get the shape of answer (C), as both rectangles have black parts. The correct answer is (C).

7. A digital clock shows the digits like this:

The clock is broken and does not show the middle horizontal line. Using the visible part of the digits we want to draw as many digits as possible. At most, how many digits can you definitely draw correctly?

(A) 10   (B) 9   (C) 8   (D) 7   (E) 6

Answer. (C)
Solution. If the clock is broken and does not show the middle horizontal line, then we cannot be sure that we can draw the digits 0 and 8 correctly. The correct answer is (C).

8. Which of the following can be equal to the sum of five consecutive whole numbers?

(A) 101   (B) 102   (C) 103   (D) 104   (E) 105

Answer. (E)
Solution. Let these five consecutive whole numbers be $n - 2$, $n - 1$, n, $n + 1$, $n + 2$, then

$$(n - 2) + (n - 1) + n + (n + 1) + (n + 2) = 5 \cdot n.$$

So, the sum of five consecutive whole numbers is divisible by 5. From these answer choices only 105 is divisible by 5. The correct answer is (E).

9. Without calculating, which of the following products is the greatest?

$$2020 \cdot 2024, \quad 2021 \cdot 2023, \quad 2019 \cdot 2025, \quad 2018 \cdot 2026, \quad 2022 \cdot 2022.$$

(A) $2020 \cdot 2024$ (B) $2021 \cdot 2023$ (C) $2019 \cdot 2025$ (D) $2018 \cdot 2026$ (E) $2022 \cdot 2022$

Answer. (E)
Solution. Recall the *difference of two squares* formula

$$(a - b) \cdot (a + b) = a^2 - b^2.$$

So, we get
$$2020 \cdot 2024 = (2022 - 2) \cdot (2022 + 2) = 2022^2 - 2^2.$$
$$2021 \cdot 2023 = (2022 - 1) \cdot (2022 + 1) = 2022^2 - 1^2.$$
$$2019 \cdot 2025 = (2022 - 3) \cdot (2022 + 3) = 2022^2 - 3^2.$$
$$2018 \cdot 2026 = (2022 - 4) \cdot (2022 + 4) = 2022^2 - 4^2.$$

So, $2022 \cdot 2022 = 2022^2$ is the greatest. The correct answer is (E).

10. Given $a - b = 2023$, where a and b are whole numbers. What is the smallest possible non-negative value of $a + b$?

(A) 0   (B) 1   (C) 20   (D) 100   (E) 2023

Answer. (B)
Solution. We want the smallest possible **non-negative** value of $a + b$.
- Can $a + b$ be 0? No, because we get $a = -b$ and as $a - b = 2023$, then $a - b = -b - b = -2b = 2023$. So $b = -\frac{2023}{2}$ and it is not a whole number.
- Can $a + b$ be 1?

$$\begin{cases} a + b = 1, \\ a - b = 2023. \end{cases}$$

Summing up both equations, we get

$$2a = 2024.$$

So, we have

$$a = 1012.$$

As $a = 1012$ and $a + b = 1$, then $b = 1 - 1012 = -1011$. So, if $a = 1012$ and $b = -1011$, then $a + b = 1$. The correct answer is (B).

Part B: Each correct answer is worth 4 points

11. The screen of a mobile phone is a 6.5×11 rectangle. There are 27 icons on the screen, so that each icon has a shape of a 1×1 square. The *available* part of the screen is the part that does not contain any icon. What is the ratio of the area of the *available* part of the screen to the area of the screen?

(A) $\frac{1}{2}$ (B) $\frac{11}{20}$ (C) $\frac{13}{24}$ (D) $\frac{89}{143}$ (E) $\frac{2}{3}$

Answer. (D)
Solution. The area of the screen is $6.5 \times 11 = 71.5$. The sum of the areas of all icons is 27. So, the area of the *available* part of the screen is $71.5 - 27 = 44.5$. Then, the ratio of the area of the *available* part of the screen to the area of the screen is:

$$\frac{71.5 - 27}{71.5} = \frac{44.5}{71.5} = \frac{445}{715} = \frac{89}{143}.$$

The correct answer is (D).

12. What is this sum equal to?

$$\frac{1}{2} + \frac{1}{3} + \frac{1}{6} + \frac{2}{3} \cdot \left(\frac{1}{2} + \frac{1}{3} + \frac{1}{6}\right) + \frac{1}{3} \cdot \left(\frac{1}{2} + \frac{1}{3} + \frac{1}{6}\right).$$

(A) 1   (B) 1.2   (C) 1.5   (D) 2   (E) 2.1

Answer. (D)
Solution. We have

$$\frac{1}{2} + \frac{1}{3} + \frac{1}{6} = \frac{3+2+1}{6} = 1.$$

So, we get that

$$\frac{1}{2} + \frac{1}{3} + \frac{1}{6} + \frac{2}{3} \cdot \left(\frac{1}{2} + \frac{1}{3} + \frac{1}{6}\right) + \frac{1}{3} \cdot \left(\frac{1}{2} + \frac{1}{3} + \frac{1}{6}\right) = 1 + \frac{2}{3} \cdot 1 + \frac{1}{3} \cdot 1 = 1 + \frac{2+1}{3} = 2.$$

The correct answer is (D).

13. How many squares are there?

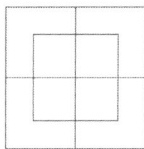

(A) 5   (B) 6   (C) 18   (D) 10   (E) 8

Answer. (D)
Solution. There are 10 squares here. The correct answer is (D).

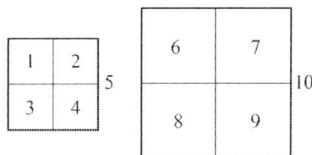

14. Little Luna has 15 red, 3 blue, and 2 orange unit cubes. She builds a tower from these cubes, placing each next cube on the previous one, so that the next cube is of a different color than the previous one. At most, how many cubes tall can Luna's tower be?

(A) 9   (B) 10   (C) 11   (D) 12   (E) 20

Answer. (C)
Solution. There are 5 cubes that are not red (3 blue and 2 orange). As she builds a tower from these cubes (placing each next cube on the previous one, so that the next cube is of a different color than the previous one), then she can use at most 6 red cubes (because there are only 5 cubes which are not red). So, the tower can be at most 11 cubes tall. Here is an example of a tower that is 11 cubes tall: *red, blue, red, blue, red, blue, red, orange, red, orange, red.*

15. Given a paper square of side length 2 and a paper circle of radius 1. Which of the following shapes is not possible to form? Cutting shapes is not allowed.

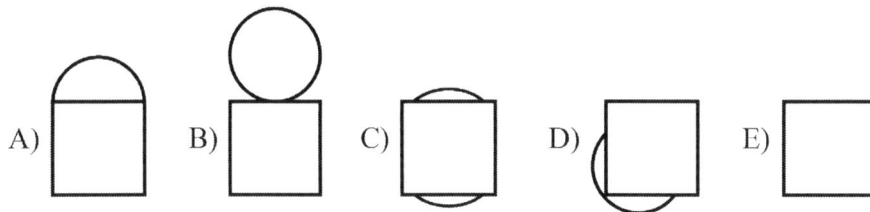

A)   B)   C)   D)   E)

Answer. (C)
Solution. The shapes of answers (A), (B). (D), (E) can be formed in the following ways.

The shape of answer (C) is not possible to form, as the length of MN will be greater than 2 (the side length of the square).

M

N

This is not possible, as the radius of the circle is 1, so MN is a diameter and must be equal to 2.

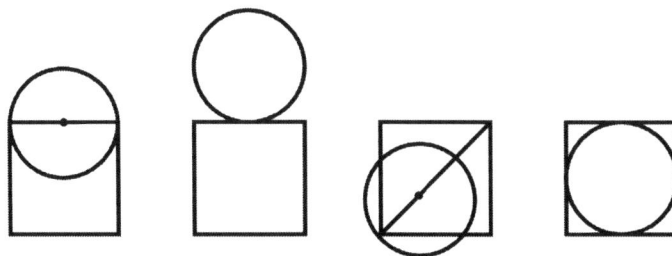

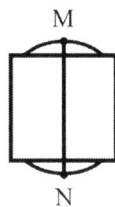

16. A big square is divided into four rectangles and one small square (see the picture). The perimeter of each rectangle is 500. What is the perimeter of the big square?

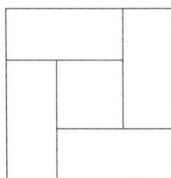

(A) 2024   (B) 1012   (C) 1000   (D) 888   (E) 1500

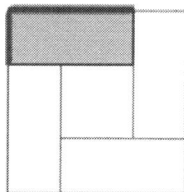

17. Given three shapes of areas a, b, c. Which of the following statements is true?

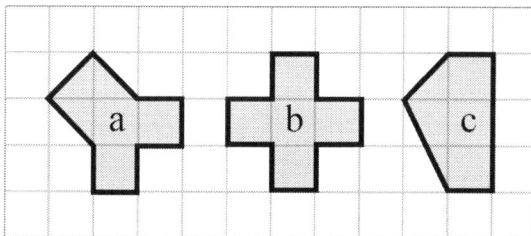

(A) $a = b$ (B) $b = c$ (C) $a > b$ (D) $a = c$ (E) $c > b$

18. Three bowls altogether contain 60 candies. The positive difference between the number of candies in any two bowls is either 4 or 8. Which of the following numbers corresponds to the number of candies in one of the bowls?

(A) 13   (B) 14   (C) 15   (D) 16   (E) 17

Answer. (D)

Solution. As the positive difference between the number of candies in any two bowls is either 4 or 8, then there are x, $x + 4$, $x + 8$ candies in these three bowls, where x is a natural number. Given that three bowls altogether contain 60 candies, then

$$x + x + 4 + x + 8 = 60.$$

We get

$$3x = 60 - 12 = 48.$$

This means

$$x = \frac{48}{3} = 16.$$

So, $x = 16$, $x + 4 = 20$, $x + 8 = 24$. Then, there are 16, 20, 24 candies in these three bowls. The correct answer is (D).

19. Five grams of paint is needed to color the surface of a cube of edge length 1 cm. Given a cube of edge length 4 cm and its eight corner cubes of edge length 1 cm are taken away. How many grams of paint is needed to color the surface of the leftover shape?

(A) 80   (B) 81   (C) 85   (D) 90   (E) 125

Answer. (A)

Solution. Note that the surface area of the leftover shape is the same as the surface area of a cube of edge length 4 cm. The surface area of a cube of edge length 4 cm is equal to the area of a 4×4 square multiplied by 6 (as a cube has 6 faces). So, it is equal to $4 \cdot 4 \cdot 6 = 96$ cm^2. Given that five grams of paint is needed to color the surface of a cube of edge length 1 cm and the surface area of a cube of edge length 1 cm is 6 cm^2 (as it has 6 faces). Note that 96 cm^2 is 16 times more than 6 cm^2, so to color a surface of 96 cm^2 one needs $16 \cdot (5 \text{ grams}) = 80$ grams of paint. The correct answer is (A).

20. A kangaroo can jump either 3 meters or 5 meters. At least, how many jumps are needed to end up exactly 29 meters away from its starting place?

(A) 8   (B) 6   (C) 10   (D) 7   (E) 9

Answer. (E)

Solution. We want to have the smallest possible number of jumps, that means that kangaroo must do as many 5 meters jumps as possible. If kangaroo does five 5 meter jumps, then it will be exactly 25 meters away from its starting place. How kangaroo can end up at a position that is exactly 4 more meters away? What is the smallest number of jumps required for that? If kangaroo does three jumps of 3 meters and one jump of 5 meters in the opposite direction, then it will be exactly $3 \cdot 3 - 5 = 4$ meters away. Then, the smallest number of jumps to be exactly 4 more meters away is four. One can easily check that with less than four jumps it is not possible. So, all together kangaroo needs to do at least $5 + 4 = 9$ jumps to end up exactly 29 meters away from its starting place. The correct answer is (E).

21. On Thursday Ann solved $\frac{1}{3}$ part of math homework exercises. On Friday she solved $\frac{1}{2}$ part of exercises left from Thursday. On Saturday she solved $\frac{1}{5}$ part of exercises left from Friday. On Sunday she solved all 4 exercises left from Saturday. How many problems did she solve on Thursday?

(A) 4   (B) 6   (C) 10   (D) 3   (E) 5

Answer. (E)

Solution. To solve this kind of problems it is easier to work backwards, that means we start from Sunday (see the figure). Given that on Sunday Ann solved all 4 exercises left from Saturday and on Saturday she solved $\frac{1}{5}$ part of exercises.

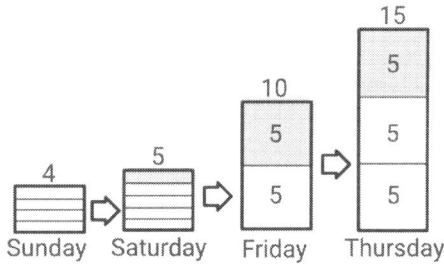

On Friday she solved $\frac{1}{2}$ part of exercises. On Thursday she solved $\frac{1}{3}$ part of exercises. So, on Thursday Ann solved 5 exercises. The correct answer is (E).

---

22. In the following Number Tower each number, except the bottom row numbers, is equal to the sum of two numbers directly below it.

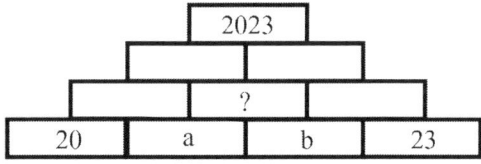

What is the value of $a + b$?

(A) 960   (B) 660   (C) 606   (D) 600   (E) 423

Answer. (B)

Solution. The numbers of the bottom row are 20, a, b, 23. Then, the numbers of the second row (counting from the bottom) are $20 + a$, $a + b$, $b + 23$. So, the numbers of the third row are $20 + 2a + b$, $23 + a + 2b$. This means

$$(20 + 2a + b) + (23 + a + 2b) = 2023.$$

So, we get

$$a + b = 660.$$

The correct answer is (B).

145

23. At least how many digits must a natural number have, so that erasing some of its digits we can get every possible two-digit number? For example, if the number is 4510 and we erase the digits 5 and 1, then we are left with the two-digit number 40.

(A) 19   (B) 17   (C) 15   (D) 14   (E) 12

> **Answer. (A)**
> **Solution.** The number must include the digits 1 and 0, so after erasing some of its digits we can get the two-digit number 10. In a similar way, it must include the digits 2, 3,..., 9.
> In order to get the two-digit number 11, the number must include the digit 1 twice. In a similar way, it must include the digits 2, 3, 4, 5, 6, 7, 8, 9 twice. So, in order to get every possible two-digit number by erasing some of its digits, it must include at least 19 digits. For example 1234567890123456789. The correct answer is (A).

24. In 2022, the grandfather had the same age as the last two digits of his birth-year, and the granddaughter had the same age as the last two digits of her birth-year. What was the sum of their ages in 2022?

(A) 70   (B) 72   (C) 75   (D) 76   (E) 80

> **Answer. (B)**
> **Solution.** Let $\overline{ab}$ be the age of the grandfather and $\overline{cd}$ be the age of the granddaughter. In 2022, the grandfather had the same age as the last two digits of his birth-year, so we get
>
> $$\overline{19ab} = 2022 - \overline{ab}.$$
>
> Then, we have
>
> $$1900 + \overline{ab} = 2022 - \overline{ab}.$$
> $$2 \cdot \overline{ab} = 2022 - 1900.$$
> $$2 \cdot \overline{ab} = 122.$$
> $$\overline{ab} = \frac{122}{2} = 61.$$
>
> So, in 2022, the grandfather was 61 years old. In 2022, the grandfather had the same age as the last two digits of her birth-year, so we get
>
> $$\overline{20cd} = 2022 - \overline{cd}.$$
>
> $$2000 + \overline{cd} = 2022 - \overline{cd}.$$
> $$2 \cdot \overline{cd} = 2022 - 2000.$$
> $$2 \cdot \overline{cd} = 22.$$
> $$\overline{cd} = \frac{22}{2} = 11.$$
>
> So, in 2022, the granddaughter was 11 years old. Then, the sum of their ages in 2022 is $61 + 11 = 72$. The correct answer is (B).

25. The greatest and the smallest of the numbers $a + b$, $b + c$, $a + c$ are equal to 62 and 61, where a, b, c are natural numbers and they can be equal to each other. What is the sum of the digits of $a + b + c$?

(A) 8 (B) 9 (C) 10 (D) 11 (E) 12

Answer. (D)

Solution. Given that the greatest and the smallest of the numbers $a+b$, $b+c$, $a+c$ are equal to 62 and 61, where a, b, c are natural numbers. So, we have three numbers, where the greatest number is 62, the smallest number is 61, and let the third number be x. This means x must be either 61 or 62, as x is a natural number and it cannot be less than 61 or greater than 62. Then, the sum of all three numbers is:

$$(a + b) + (b + c) + (a + c) = 62 + 61 + x.$$

We get

$$2 \cdot (a + b + c) = 123 + x.$$

The left side of this equation is an even number, so the right side must also be an even number. This means $x = 61$. Then

$$2 \cdot (a + b + c) = 123 + 61 = 184.$$

We get

$$a + b + c = 92.$$

So, the sum of the digits is $9 + 2 = 11$. The correct answer is (D).

26. Little Luna picked two different digits and kept them secret from her grandmother. Luna said the sum of these two digits to her grandmother, after that the grandmother said that she for sure knows both digits. What is the value of one of these two digits?

(A) 9 (B) 8 (C) 1 or 8 (D) 0 (E) 0 or 9

Answer. (E)

Solution. The sum of two different digits can be any number from 1 to 17. If the sum of these two digits is any of the following numbers 3, 4,..., 15, then the grandmother cannot for sure know both digits, as

$$3 = 3 + 0, \quad 3 = 1 + 2,$$

$$4 = 4 + 0, \quad 4 = 1 + 3,$$

$$\ldots$$

$$15 = 7 + 8, \quad 15 = 9 + 6.$$

So, Luna said to her grandmother one of the following numbers 1, 2, 16, 17. Then, these two digits must be either 0,1, or 0,2, or 7,9, or 8,9. So, one of the digits must be either 0 or 9. The correct answer is (E).

27. There are 28 domino tiles, so that the height of each tile is 0.4 cm and the width is 1 cm. We want to fit all tiles into a rectangular box, so that 28 tiles are placed in 4 layers and each layer includes 7 tiles. Tiles can be placed in each layer in both of the following ways:

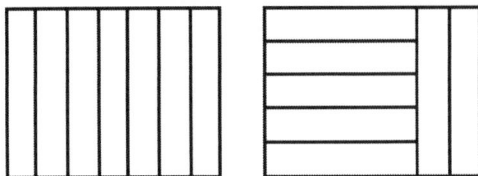

What is the smallest possible volume of such a box?

(A) 59 cm^3 (B) 58 cm^3 (C) 57 cm^3 (D) 56 cm^3 (E) 55 cm^3

Answer. (D)

Solution. The length of each tile is equal to 5 cm.

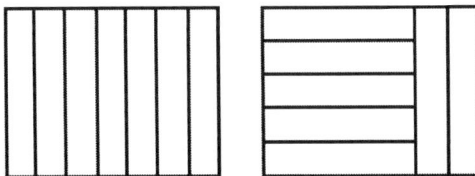

So, the volume of each tile is $5 \cdot 1 \cdot 0.4 = 2$ cm^3. Then, the volume of all 28 tiles is equal to $28 \cdot 2$ cm$^3 = 56$ cm^3. So, the smallest possible volume of such a box can possibly be 56 cm^3. As there are 4 layers in the box and the height of each tile is 0.4 cm, then the total height of the box must be 1.6 cm. The length of each tile is 5 cm. Let us divide 56 by $1.6 \cdot 5$ to find out what the width of the box must be. Note that

$$\frac{56}{1.6 \cdot 5} = 8.$$

So, the width of the box must be 8 cm.

Then, the smallest possible volume of such a box is 56 cm^3. The correct answer is (D).

28. a and b are the remainders when natural numbers n and $4n$ are divided by 10 and 15, respectively. What is the greatest possible value of $a + b$?

(A) 19   (B) 20   (C) 21   (D) 22   (E) 23

Answer. (B)
Solution. As a is the remainder after division by 10, then $0 \le a \le 9$. As b is the remainder after division by 15, then $0 \le b \le 14$. So, we get

$$a + b \le 23.$$

As a is the remainder when natural number n is divided by 10, then

$$n = 10 \cdot q + a,$$

As b is the remainder when natural number $4n$ is divided by 15, then

$$n = 15 \cdot k + b,$$

where the division quotients q and k are natural numbers. Adding the last two equations, we get
$$n + 4n = 10 \cdot q + 15 \cdot k + a + b = 5(2 \cdot q + 3 \cdot k) + a + b.$$

Note that $n + 4n = 5n$ and $5(2 \cdot q + 3 \cdot k)$ are divisible by 5, then $a + b$ must be divisible by 5 as well. We have $a + b \le 23$ and $a + b$ is divisible by 5. So, the greatest possible value of $a + b$ may be 20. If we choose $a = 7$ and $b = 13$, then $a + b = 20$. The correct answer is (B).

29. Several pairs of natural numbers add up to 77. For example, 1 and 76, or 32 and 45. From all these pairs that add up to 77 we choose one pair of numbers that has the smallest possible LCM (least common multiple). What is the sum of the digits of their LCM?

(A) 10   (B) 11   (C) 12   (D) 13   (E) 15

Answer. (C)
Solution. Let (a, b) be the pair so that $a + b = 77$ and $\text{LCM}(a, b)$ is the smallest possible. Denote $\text{LCM}(a, b) = n$, then a and b are factors of n and $a \ne b$. This means that

$$77 = a + b \le n + \frac{n}{6}.$$

We get $n \ge 66$. We provide an example where n can be 66. For example, $77 = 66 + 11$ and $\text{LCM}(66, 11) = 66$. So, the smallest possible LCM is 66 and its sum of the digits is $6 + 6 = 12$. The correct answer is (C).

30. The shaded rectangle is placed on a rectangle of side length 3 (see the figure). What is the ratio of the area of the rectangle of side length 3 to the area of the shaded rectangle?

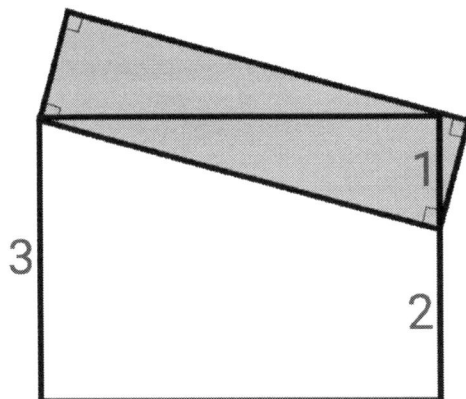

(A) 2:1   (B) 5:2   (C) 9:4   (D) 11:4   (E) 3:1

Answer. (E)

Solution. Consider the picture below.

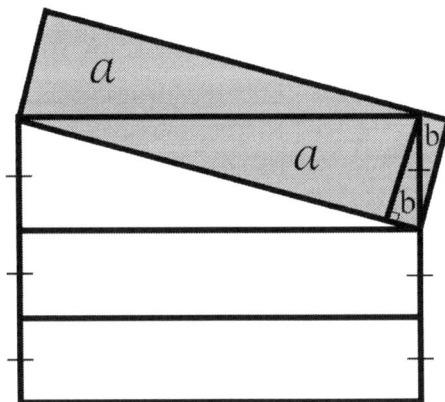

The area of the shaded rectangle is equal to $a + a + b + b = 2 \cdot (a + b)$. The area of the bold rectangle of side length 1 is equal to $(a + b) + (a + b) = 2 \cdot (a + b)$. The area of the rectangle of side length 3 is three times more than the area of the bold rectangle of side length 1 (because one of their side lengths is the same and the other side is three times more), so the area of the rectangle of side length 3 is $3 \cdot 2 \cdot (a + b) = 6 \cdot (a + b)$. Then, the ratio of the area of this rectangle (of side length 3) to the area of the shaded rectangle is $\left(6 \cdot (a + b)\right) : \left(2 \cdot (a + b)\right) = 3 : 1$. The correct answer is (E).

Solutions of Test 7

Part A: Each correct answer is worth 3 points

1. At most how many days can there be in six consecutive months?

(A) 181   (B) 182   (C) 183   (D) 184   (E) 185

> **Answer. (D)**
> **Solution.** Six months is half of the year and a year has either 365 days or 366 days. So, it looks like that six months need to have at most 183 days. On the other hand, July and August are two consecutive months that both have the greatest possible number of days (31 days). So, we can take for example July (31 days), August (31 days), September (30 days), October (31 days), November (30 days), December (31 days). Then, we get that these six consecutive months have $31 + 31 + 30 + 31 + 30 + 31 = 184$ days. One can easily check that any six consecutive months cannot have more than 184 days. The correct answer is (D).

2. What is the coordinate of point A?

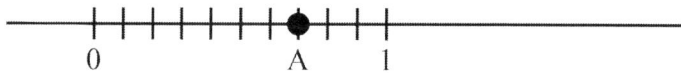

(A) 0.5   (B) 0.6   (C) 0.7   (D) 0.8   (E) 0.9

> **Answer. (C)**
> **Solution.** The length of this line segment is 1 and it is divided into 10 equal parts. So, the coordinate of point A is $\frac{7}{10} = 0.7$. The correct answer is (C).

3. For the digit x we know that $3.0023 < 3.x023 < 3.6023$. What is the sum of all possible values of the digit x?

(A) 10   (B) 11   (C) 12   (D) 15   (E) 16

> **Answer. (D)**
> **Solution.** As $3.0023 < 3.x023 < 3.6023$, then all possible values of the digit x are 1, 2, 3, 4, 5. So, the sum of all possible values of x is $1 + 2 + 3 + 4 + 5 = 15$. The correct answer is (D).

4. Math operations $\triangle$ and $\square$ are defined in the following way: $a \triangle b = a + b + 1$ and $a \square b = (a-1) \cdot b$. What is $(1.1 \triangle 1.4) \square 2.6$?

(A) 6.5   (B) 4.5   (C) 4.3   (D) 4.2   (E) 3.6

Answer. (A)
Solution. We have
$$1.1 \triangle 1.4 = 1.1 + 1.4 + 1 = 3.5.$$
So, we get
$$(1.1 \triangle 1.4) \square 2.6 = 3.5 \square 2.6 = (3.5 - 1) \cdot 2.6 = 2.5 \cdot 2.6 = 6.5.$$
The correct answer is (A).

5. Which picture shows $\angle BCA$?

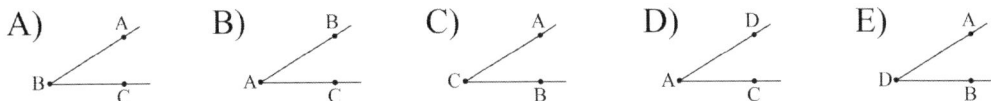

Answer. (C)
Solution. The correct answer is (C). For example, (A) shows the angle $\angle ABC$ (or $\angle CBA$).

6. Some squares of a 6×6 square are numbered by numbers 1, 2,..., 18. The paper was folded across line l, then the paper was folded across line m. After this, with which square will the square number 4 coincide?

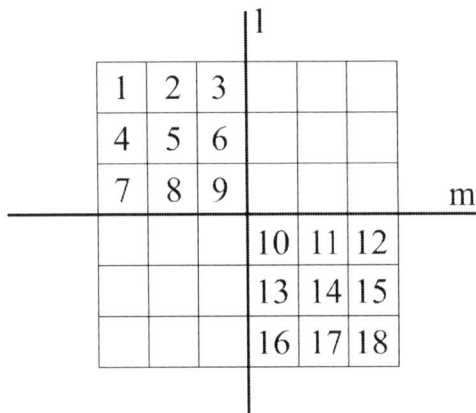

(A) 12   (B) 11   (C) 18   (D) 17   (E) 15

Answer. (E)

Solution. When the paper is folded across the line l, then the square number 4 will coincide with the ★ square (see the figure).

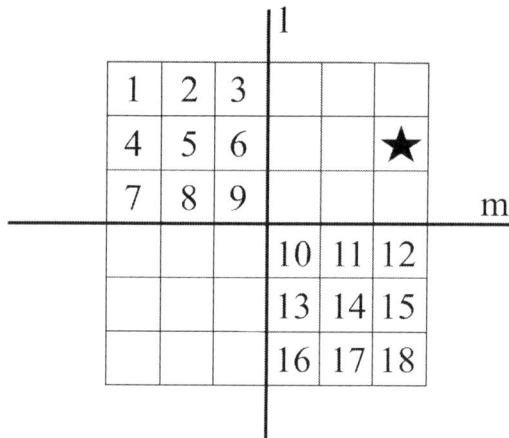

When the paper is folded across the line m, then the ⋆ square will coincide with the square number 15. The correct answer is (E).

---

7. A local supermarket increases the price of its vanilla ice-cream from \$5 by 20 percent. At most, how many vanilla ice-creams can you buy with \$50?

(A) 10   (B) 9   (C) 8   (D) 7   (E) 6

Answer. (C)

Solution. Given that the price of vanilla ice-cream was \$5 and it increased by 20 percent, then the new price will be

$$\$5 + \frac{\$5 \cdot 20}{100} = \$6.$$

So, you can buy at most 8 vanilla ice-creams (because $6 \cdot 8 = 48 < 50$ and $6 \cdot 9 = 54 > 50$). The correct answer is (C).

---

8. A woodcutter needs 7.5 minutes to cut a 6 meters long wooden stick into 6 one meter long wooden sticks. How many minutes are needed to cut a 11 meters long wooden stick into 11 one meter long wooden sticks?

(A) 11.5   (B) 12.5   (C) 13   (D) 15   (E) 15.5

Answer. (D)

Solution. In order to cut a 6 meters long wooden stick into 6 one meter long wooden sticks the woodcutter needs to cut it in 5 different places. So for each cut the woodcutter needs $\frac{7.5 \text{ minutes}}{5} = 1.5$ minutes. In order to cut a 11 meters long wooden stick into 11 one meter long wooden sticks the woodcutter needs to cut it in 10 different places. As for each cut 1.5 minutes are needed, then for 10 cuts 15 minutes are needed. The correct answer is (D).

9. At least how many different digits must be used in order to write three consecutive three-digit numbers?

(A) 2   (B) 3   (C) 4   (D) 5   (E) 6

Answer. (B)
Solution. If we write three consecutive three-digit numbers, then they will have different last digits. This means that we need to use at least three digits. Here is an example of three consecutive three-digit numbers, so that only three digits are used (only the digits 0, 1, and 2 are used): 100, 101, 102. The correct answer is (D).

10. Given five different paper shapes (see the figure). Which two of these shapes can be put together to form a circle? Papers cannot be flipped.

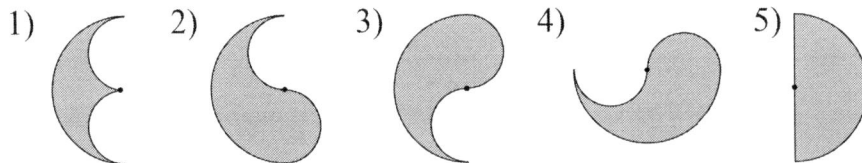

(A) 1 and 2   (B) 2 and 3   (C) 3 and 4   (D) 4 and 5   (E) 2 and 4

Answer. (E)
Solution. The correct answer is (E), see the figure.

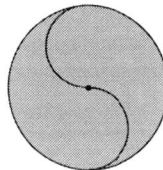

154

Part B: Each correct answer is worth 4 points

11. The sum of perimeters of triangle ABC and triangle PQR is equal to 20. What is the sum of perimeters of the shaded six triangles?

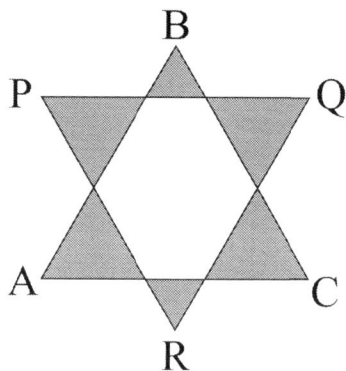

(A) 10   (B) 20   (C) 40   (D) 45   (E) 50

Answer. (B)
Solution. Note that the sum of side lengths of all shaded six triangles is equal to the sum of side lengths of triangle ABC and triangle PQR. As the sum of perimeters of triangle ABC and triangle PQR is equal to 20, then the sum of perimeters of the shaded six triangles is equal to 20. The correct answer is (B).

12. At most how many odd digits can the product of two two-digit numbers have?

(A) 0   (B) 1   (C) 2   (D) 3   (E) 4

Answer. (E)
Solution. The product of two two-digit numbers is less than $100 \cdot 100$. This means that the product of two two-digit numbers can be at most a four-digit number. Can all four digits be odd numbers? Yes, for example $91 \cdot 83 = 7553$. The correct answer is (E).

13. Two cars started simultaneously toward each other from towns A and B, and they met in 1 hour. The car which came from A turned back and drove back to A, and the other car continued its way to A. Given that the car which came from A reached A 20 minutes earlier than the other car. How much time did the car which started from B spend on the whole trip from B to A?

(A) 1 hour 20 mins   (B) 2 hours 20 mins   (C) 2 hours 30 mins   (D) 3 hours   (E) 4 hours

Answer. (B)
Solution. As the cars met in 1 hour and as the car which came from A turned back and drove back to A, this means that this car has spend 1 hour for going back. So, the car which came from A has spent 2 hours to get back to A. Given that the car which came from A reached A 20 minutes earlier than the other car, this means that the car which came from B has spent 2 hours and 20 minutes on the whole trip from B to A. The correct answer is (B).

14. In how many different ways is possible to write the numbers 1, 2, 3, 4 in the squares of a 2×2 square, so that the sum of the numbers of any two squares that share a side is a prime number?

(A) 8   (B) 10   (C) 16   (D) 20   (E) 24

Answer. (A)
Solution. Let us color this 2×2 square in the following way (see the figure).

Note that 2 and 4 must be written in the squares of the same color, because $2 + 4 = 6$ is a composite number. So, the number of all possible ways is equal to $2 \cdot 2 \cdot 2 = 8$. Then, the sums of the numbers of any two squares that share a side are 3, 5, 5, 7. So, they are prime numbers. The correct answer is (A).

15. A rectangle is divided into nine squares (see the figure), so that the side length of the smallest square is equal to 1. What is the side length of the greatest side of the rectangle?

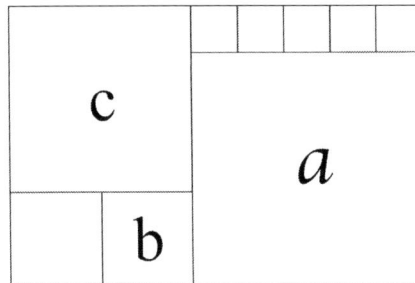

(A) 6   (B) 7   (C) 8   (D) 9   (E) 10

Answer. (D)
Solution. Note that the side length of square A is $5 \cdot 1 = 5$ (see the figure), so the side length of the smallest side of the rectangle is $1 + 5 = 6$. Note that the side length of square C is twice more than the side length of square B, and their sum is 6. So, the side length of square B is $\frac{6}{3} = 2$. Then, the side length of the greatest side of the rectangle is $5 + 2 + 2 = 9$. The correct answer is (D).

16. In each square of a 3×3 is written a number (see the figure). The places of two numbers were switched, so that the product of all numbers of the shaded squares is equal to the product of all numbers of the white squares. What is the sum of these two numbers?

6	5	1
2	7	10
14	9	12

(A) 12   (B) 17   (C) 19   (D) 21   (E) 24

> **Answer. (C)**
> **Solution.** Note that the product of all numbers of the shaded squares is equal to $2^2 \cdot 3^2 \cdot 7^2$ and the product of all numbers of white squares is equal to $2^4 \cdot 3^2 \cdot 5^2$. In order these products to be equal we need to switch $2 \cdot 5$ and 7 (then both products will be equal to $2^3 \cdot 3^2 \cdot 5 \cdot 7$). So, in order the product of all numbers of the shaded squares to be equal to the product of all numbers of the white squares we need to switch 10 and 7. Then, the sum of these two numbers is $10 + 7 = 17$. The correct answer is (B).

17. A toy kangaroo is on the bottom leftmost square of a chessboard. It can only jump horizontally or vertically either 3 squares or 2 squares. At least, how many jumps are needed to get to the top rightmost square?

(A) 3   (B) 4   (C) 5   (D) 6   (E) 7

> **Answer. (D)**
> **Solution.** It must move 7 squares right and 7 squares up. In order to move 7 squares right it must do 3 jumps (3 squares, 3 squares, then 2 squares), in order to move 7 squares up it must do 3 jumps (3 squares, 3 squares, then 2 squares). So, all together it must do 6 jumps. The correct answer is (D).

18. A rectangle of whole number side lengths is divided into two rectangles of whole number side lengths (see the figure), so that the perimeters of the small rectangles are 20 and 24. What is the smallest possible perimeter of the initial rectangle?

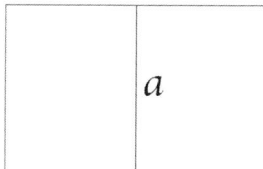

(A) 20   (B) 22   (C) 24   (D) 26   (E) 44

Answer. (D)

Solution. As the perimeters of the smaller rectangles are 20 and 24, then the perimeter of the initial rectangle is $20+24-2 \cdot a$. So, the perimeter of the initial rectangle is the smallest possible when $44 - 2 \cdot a$ is the smallest possible. Note that $44 - 2 \cdot a$ is the smallest possible when a is the largest possible. As the perimeter of one of the small rectangles is $2a + 2b = 20$ and a, b are positive whole numbers, then the largest possible value of a is 9. Then, $44 - 2a = 44 - 2 \cdot 9 = 26$. So, the smallest possible perimeter of the initial rectangle is 26. The correct answer is (D).

19. How many rectangles are there? (see the figure, a square is also a rectangle).

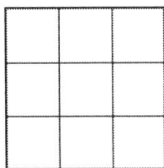

(A) 27 (B) 30 (C) 32 (D) 35 (E) 36

Answer. (E)

Solution. The possible sizes of all possible rectangles are 1×1, 1×2, 1×3, 2×2, 2×3, 3×3.
1×1 rectangles: there are 9 of them.
1×2 rectangles: there are $2 \cdot 2 \cdot 3 = 12$ of them.
1×3 rectangles: there are 6 of them.
2×2 rectangles: there are 4 of them.
2×3 rectangles: there are 4 of them.
3×3 rectangles: there are 1 of them.
So, altogether there are $9 + 12 + 6 + 4 + 4 + 1 = 36$ rectangles. The correct answer is (E).

20. From the list of the numbers 1, 2,..., 100 all numbers divisible by 2 are erased, then all numbers divisible by 3 are erased, then all numbers divisible by 4 are erased (if there are any left), and so on. What is the product of the digits of the last erased number?

(A) 10 (B) 20 (C) 30 (D) 62 (E) 63

Answer. (E)

Solution. Note that the last erased number is the greatest prime number that is smaller than 100, so it is 97. So, the product of its digits is $9 \cdot 7 = 63$. The correct answer is (E).

Part C: Each correct answer is worth 5 points

21. In how many different ways is possible to cut (by a straight line) a 4×6 paper rectangle into two parts, so that when one part is folded it becomes the same as the other part?

(A) 0   (B) 1   (C) 2   (D) 3   (E) 4

Answer. (C)

Solution. If these two parts are the same, then their areas must be equal. So, the straight line that splits this 4×6 rectangle into two equal parts must pass through the intersection point of the diagonals. So, the following two figures are possible.

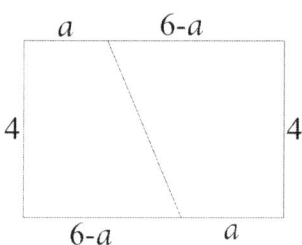

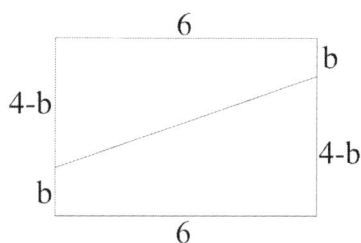

In the first case, we get $a = 6 - a$. Then $a = 3$.
In the second case, we get $b = 4 - b$. Then $b = 2$.
So, there are two possible ways to cut by a straight line a 4×6 paper rectangle into two parts that are the same. The correct answer is (C).

22. What change in percent is made to the area of a rectangle by decreasing its length and its perimeter by 10 percent?

(A) 25   (B) 26   (C) 20   (D) 19   (E) 21

Answer. (D)

Solution. Let the length of the rectangle be 100 units and its width be x units, then its area is $100 \cdot x$ and its perimeter is $100 + 100 + x + x = 200 + 2x$. By decreasing the length of the rectangle by 10 percent the length of the new rectangle is:

$$100 - \frac{100 \cdot 10}{100} = 90.$$

Decreasing the perimeter of the rectangle by 10 percent the perimeter of the new rectangle is:

$$(200 + 2x) - \frac{(200 + 2x) \cdot 10}{100} = (200 + 2x) - \frac{(200 + 2x)}{10} =$$

$$= (200 + 2x) - (20 + 0.2x) = 180 + 1.8x.$$

This means that the width of the new rectangle is:

$$\frac{180 + 1.8x - 90 - 90}{2} = 0.9x.$$

Then, the area of the new rectangle is:

$$90 \cdot 0.9x = 81x.$$

The area of the rectangle was $100x$, and $81x$ is smaller than $100x$ by 19 percent. The correct answer is (D).

23. Two cyclists start simultaneously toward each other from cities A and B. One going at 14 miles per hour. Given that when they were 10 miles apart, 40 minutes after that they were again 10 miles apart. What is the speed (in miles per hour) of the other cyclist?

(A) 12   (B) 14   (C) 15   (D) 16   (E) 17

Answer. (D)

Solution. Given that when they were 10 miles apart, 40 minutes after that they were again 10 miles apart. This means that in 40 minutes they covered a distance of 20 miles. Note that 40 minutes is $\frac{2}{3}$ hours. So, in order to cover 20 miles in $\frac{2}{3}$ hours the speed must be $\frac{20}{\frac{2}{3}} = 30$ miles per hour. So, the sum of their speeds must be 30 miles per hour. As the speed of one of them is 14 miles per hour, then the speed of the other one is $30 - 14 = 16$ miles per hour. The correct answer is (D).

24. At least how many numbers must be added to the list of the numbers 15, 20, 80, so that the positive difference of any two of the numbers belongs to the new list of these numbers?

(A) 1   (B) 5   (C) 10   (D) 12   (E) 13

Answer. (E)

Solution. Given that the positive difference of any two of the numbers needs to belong to the new list of these numbers, then

20 - 15 = 5, so 5 must be added to the list.
15 - 5 = 10, so 10 must be added to the list.
80 - 5 = 75, so 75 must be added to the list.
75 - 5 = 70, so 70 must be added to the list.
70 - 5 = 65, so 65 must be added to the list.
65 - 5 = 60, so 60 must be added to the list.
60 - 5 = 55, so 55 must be added to the list.
55 - 5 = 50, so 50 must be added to the list.
50 - 5 = 45, so 45 must be added to the list.
45 - 5 = 40, so 40 must be added to the list.
40 - 5 = 35, so 35 must be added to the list.
35 - 5 = 30, so 30 must be added to the list.
30 - 5 = 25, so 25 must be added to the list.

Note that the list of the numbers 5, 10, 15,..., 80 satisfy the conditions of the problem. There are 16 numbers in this list. So, 13 numbers need to be added to the list of the numbers 15, 20, 80, so that the positive difference of any two of the numbers belongs to the new list of these numbers. The correct answer is (E).

---

25. A three-digit number is called "beautiful" if the difference of that three-digit number and the product of its digits is equal to 110. How many "beautiful" three-digit numbers are there?

(A) 8   (B) 10   (C) 9   (D) 12   (E) 11

Answer. (B)

Solution. Consider a three-digit number $\overline{abc}$, then given that

$$110 = \overline{abc} - a \cdot b \cdot c.$$

So, we get

$$110 = \overline{abc} - a \cdot b \cdot c = 100 \cdot a + 10 \cdot b + c - a \cdot b \cdot c = a \cdot (100 - b \cdot c) + 10 \cdot b + c \geq$$

$$\geq 1 \cdot (100 - b \cdot c) + 10 \cdot b + c = 100 + b \cdot (10 - c) + c \geq 100 + 1 \cdot (10 - c) + c = 110.$$

The equality holds true when $a = 1$, $b = 1$, and c is any digit. So, all possible such numbers are 110, 111, 112, 113, 114, 115, 116, 117, 118, 119. This mean that there are 10 "beautiful" three-digit numbers. The correct answer is (B).

---

26. In how many different ways is possible to divide the numbers 1, 2, 3, 4, 5, 6, 7, 8 into four pairs, so that the sum of the numbers in any pair is a prime number?

(A) 6   (B) 7   (C) 8   (D) 9   (E) 10

Answer. (A)

Solution. Note that the numbers 2, 4, 6, 8 must be in different pairs. Otherwise, if any two of them is in the same pair than their sum will be an even number greater than 2 (so it cannot be a prime number).

8 can be in the same pair with 3 or 5
2 can be in the same pair with 1, 3, or 5
4 can be in the same pair with 1, 3, or 7
6 can be in the same pair with 1, 5, or 7
So, all possible such pairs are the following pairs:

$$(8,3), (2,1), (4,7), (6,5).$$

$$(8,3), (2,5), (4,1), (6,7).$$

$$(8,3), (2,5), (4,7), (6,1).$$

$$(8,5), (2,1), (4,3), (6,7).$$

$$(8,5), (2,3), (4,1), (6,7).$$

$$(8,5), (2,3), (4,7), (6,1).$$

So, the number of all possible different ways is 6. The correct answer is (A).

---

27. a, b, c, d is a rearrangement of the numbers 1, 2, 3, 4, so that:
a is divisible by b,
$a + b$ is divisible by c,
$a + b + c$ is divisible by d.
How many such rearrangements are there?

(A) 0   (B) 1   (C) 2   (D) 3   (E) 4

Answer. (C)
Solution. Given that $a + b + c$ is divisible by d, then $a + b + c + d$ is also divisible by d. As $a + b + c + d = 1 + 2 + 3 + 4 = 10$, then we have that 10 is divisible by d. So, $d = 1$ or $d = 2$.
Case 1. If $d = 1$. Given that $a + b$ is divisible by c, then $a + b + c$ is also divisible by c. As $d = 1$ and $a + b + c + d = 10$, then $a + b + c = 9$. So, 9 is divisible by c. Then $c = 3$. Given that a is divisible by b, then $a = 4$ and $b = 2$.
So, in this case we get this rearrangement: 4, 2, 3, 1.
Case 2. If $d = 2$. Given that $a + b$ is divisible by c, then $a + b + c$ is also divisible by c. As $d = 2$ and $a + b + c + d = 10$, then $a + b + c = 8$. So, 8 is divisible by c. Then $c = 1$ or $c = 4$. Note that the case $c = 1$ is not possible, because in that case we get $a = 3$, $b = 4$ or $a = 4$, $b = 3$ (so the condition a is divisible by b is not correct for both of these cases).
For $c = 4$, as a is divisible by b, then we get $a = 3$ and $b = 1$.
So, in this case we get this rearrangement: 3, 1, 4, 2.
This means that, there are two such rearrangements (one per each case, namely 4, 2, 3, 1 and 3, 1, 4, 2). The correct answer is (C).

28. At least how many digits must a natural number have, so that erasing some of its digits we can get every possible two-digit number which is divisible by 3? For example, if the number is 9239 and we erase the digits 2 and 3, then we are left with the two-digit number 99.

(A) 15   (B) 16   (C) 17   (D) 18   (E) 19

Answer. (B)
Solution. Examples of two-digit numbers that are divisible by 3 are:

$$30, 33, 66, 99.$$

So, the natural number we are looking for must include the digits 0, 3, 3, 6, 6, 9, 9. There are seven digits listed here.
Examples of two-digit numbers that are divisible by 3 are:

$$18, 81, 48, 84, 78, 87.$$

So, the natural number we are looking for must include the digit 8 and the digits 1, 4, 7 must appear from both left and right sides of the digit 8. So, it must include the digits 8, 1, 1, 4, 4, 7, 7. There are seven digits listed here.
Examples of two-digit numbers that are divisible by 3 are:

$$27, 51.$$

So, the natural number we are looking for must include the digits 2 and 5 as well. There are two digits listed here.
We get that the natural number we are looking for must include at least $7 + 7 + 2 = 16$ digits. For example, this number that has 16 digits satisfies the conditions of the problem.

$$1346790258134679.$$

So, it must have at least 16 digits. The correct answer is (B).

163

29. a, b, c, d are different positive digits. Eleven two-digit numbers were chosen from the following twelve two-digit numbers $\overline{ab}$, $\overline{ac}$, $\overline{ad}$, $\overline{bc}$, $\overline{ba}$, $\overline{bd}$, $\overline{ca}$, $\overline{cb}$, $\overline{cd}$, $\overline{da}$, $\overline{db}$, $\overline{dc}$, so that the sum of chosen numbers is equal to 450. What is the sum of the digits of the two-digit number that was not chosen?

(A) 3   (B) 4   (C) 5   (D) 6   (E) 7

Answer. (A)

Solution. Note that the sum of all twelve numbers is equal to

$$\overline{ab} + \overline{ac} + \overline{ad} + \overline{bc} + \overline{ba} + \overline{bd} + \overline{ca} + \overline{cb} + \overline{cd} + \overline{da} + \overline{db} + \overline{dc} = (10 \cdot a + b) + (10 \cdot a + c) +$$

$$+ (10 \cdot a + d) + (10 \cdot b + c) + (10 \cdot b + a) + (10 \cdot b + d) + (10 \cdot c + a) + (10 \cdot c + b) +$$

$$+ (10 \cdot c + d) + (10 \cdot d + a) + (10 \cdot d + b) + (10 \cdot d + c) = 33 \cdot (a + b + c + d).$$

So, the sum of all twelve numbers is divisible by 33. If we denote "the number that was not chosen" by x, then this sum is equal to $450 + x$. As x is a two-digit number (greater than 9 and smaller than 100), then $459 < 450 + x < 550$. The numbers between 459 and 550 that are divisible by 33 are $462 = 33 \cdot 14$, $495 = 33 \cdot 15$, and $528 = 33 \cdot 16$. So, x can be equal to $462 - 450 = 12$, $495 - 450 = 45$, or $528 - 450 = 78$.

If $x = 12$, then $450 + 12 = 33 \cdot (a + b + c + d)$.
We get $a + b + c + d = \frac{462}{33} = 14$. As $x = 12$ and as $a + b + c + d = 14$, then the sum of the other two digits is equal to $14 - (1 + 2) = 11$. We can take for example 3 and 8, as $3 + 8 = 11$.

If $x = 45$, then $450 + 45 = 33 \cdot (a + b + c + d)$.
We get $a + b + c + d = \frac{495}{33} = 15$. As $x = 45$ and as $a + b + c + d = 15$, then the sum of the other two digits is equal to $15 - (4 + 5) = 6$. As a, b, c, d must be different digits and 4 and 5 are already used, then it is not possible to choose two digits, so that their sum is equal to 6.

If $x = 78$, then $450 + 78 = 33 \cdot (a + b + c + d)$.
We get $a + b + c + d = \frac{528}{33} = 16$. As $x = 78$ and as $a + b + c + d = 16$, then the sum of the other two digits is equal to $16 - (7 + 8) = 1$. This is not possible, as a, b, c, d are different positive digits, so the sum of any two of them is greater than or equal to $1 + 2 = 3$.

164

30. The numbers 1, 2, 3, 4, 5, 6 are written on the faces of a cube (one number per face). For any two faces that share a common edge the positive difference of the numbers written on these two faces is written on that edge. What is the greatest possible sum of all numbers written on all 12 edges?

(A) 30   (B) 31   (C) 32   (D) 33   (E) 34

Answer. (C)

Solution. Let this sum be S. If we replace 1 by 2, then the sum S will be replaced by $S - 4$. If we replace 6 by 5, then $S - 4$ will be replaced by $S - 8$. This means that the new numbers written on the faces of the cube are 2, 2, 3, 4, 5, 5, and the sum of all numbers written on all 12 edges is $S - 8$. Note that from these 12 numbers at most four of them can be equal to 3, at most four of them can be equal to 2, and at most four of them can be equal to 1. Then, we get

$$S - 8 \leq 4 \cdot 3 + 4 \cdot 2 + 4 \cdot 1.$$

So, we get $S \leq 32$. We provide an example for $S = 32$.

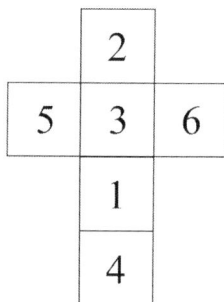

	2	
5	3	6
	1	
	4	

The correct answer is (C).

Bibliography

[1] Sedrakyan H., Sedrakyan N., *AMC 8 preparation book*, USA (2021)

[2] Sedrakyan H., Sedrakyan N., *Competition math for middle school: must-knows and beyond*, USA (2023)

[3] Sedrakyan H., Sedrakyan N., *AMC 10 preparation book*, USA (2021)

[4] Sedrakyan H., Sedrakyan N., *AMC 12 preparation book*, USA (2021)

[5] Sedrakyan H., Sedrakyan N., *AIME preparation book*, USA (2022)

[6] Sedrakyan H., Sedrakyan N., *AMC and AIME geometry: must-know techniques*, USA (2023)

[7] Sedrakyan H., Sedrakyan N., *Number theory through exercises*, USA (2019)

[8] Math Kangaroo, International Competition in Mathematics

Made in the USA
Las Vegas, NV
28 May 2024